AF831084

Pullulan

Pullulan

Processing, Properties, and Applications

edited by

Shakeel Ahmed

Aisverya Soundararajan

JENNY STANFORD
PUBLISHING

Published by

Jenny Stanford Publishing Pte. Ltd.
Level 34, Centennial Tower
3 Temasek Avenue
Singapore 039190

Email: editorial@jennystanford.com
Web: www.jennystanford.com

British Library Cataloguing-in-Publication Data
A catalogue record for this book is available from the British Library.

Pullulan: Processing, Properties, and Applications

Copyright © 2021 by Jenny Stanford Publishing Pte. Ltd.
All rights reserved. This book, or parts thereof, may not be reproduced in any form or by any means, electronic or mechanical, including photocopying, recording or any information storage and retrieval system now known or to be invented, without written permission from the publisher.

For photocopying of material in this volume, please pay a copying fee through the Copyright Clearance Center, Inc., 222 Rosewood Drive, Danvers, MA 01923, USA. In this case permission to photocopy is not required from the publisher.

ISBN 978-981-4877-17-6 (Hardcover)
ISBN 978-1-003-05694-2 (eBook)

Contents

Preface

Pullulan is a polysaccharide that consists of maltotriose units connected by α-1,4 glycosidic bonds. It is produced by fungus *Aureobasidium pullulans* grown on starch. Pullulans have a wide range of commercial and industrial applications, such as in health care, food science, and pharmaceuticals. Properties such as excellent transparent film-forming ability, moisture absorptivity, water solubility, and non-toxicity have made pullulan advantageous as an ingredient in the manufacturing of cosmetics. Due to its high water solubility, it serves as an ideal ingredient for eye care, bathing, and hair products. Due to its adhesive nature, it is chiefly used as a film former for oral, oxygen-impermeable biofilms, as well as in the formation of fibers and compression mouldings. Pullulan is widely exploited in the fields of tissue engineering and grafting as well as medical imaging. It also acts as a drug carrier, plasma volume expander, and molecular chaperone.

This book discusses the chemistry, properties, processing, importance, and versatile applications of pullulan, along with its production at the laboratory level. It comprehensively summarizes many of the recent technical research findings related to this polysaccharide. This is a unique book with contributions from the experts working in this research domain and will be a useful reference for scientists, academicians, research scholars, chemists, and technologists.

We hope this book will assist readers at graduate and post-graduate level in academics and research.

Shakeel Ahmed

Aisverya Soundararajan

Autumn 2020

Chapter 1

Polysaccharides: An Overview

S. Vijayanand,[a] Ashwini Ravi,[a] Aisverya Soundararajan,[b] Annu,[c] P. N. Sudha,[b] and J. Hemapriya[d]

[a]*Bioresource Technology Lab, Department of Biotechnology, Thiruvalluvar University, Serkkadu, Vellore 632115, Tamil Nadu, India*
[b]*Biomaterial Research Lab, Department of Chemistry, DKM College for Women (Autonomous), Vellore 632001, Tamil Nadu, India*
[c]*Department of Chemistry, Jamia Millia Islamia, New Delhi 110025, India*
[d]*PG & Research Department of Microbiology, DKM College for Women (Autonomous), Vellore 632001, Tamil Nadu, India*
micropriya80@gmail.com

Polysaccharides, commonly known as "Cinderella of biopolymers," are molecules composed of monosaccharide residues, which are joined together by O-glycosidic linkages. Naturally, polysaccharides are produced by various biological entities such as plants, microbes, seaweeds, and animals. Polysaccharides were often found to have linear structure, and it may also contain various degrees of branching. Despite their structure, they also vary in various components such as monosaccharide composition, linkage type, pattern of linkage, chain type, degree of polymerization, etc. based on the source of production. This variable structure of polysaccharides is responsible

Pullulan: Processing, Properties, and Applications
Edited by Shakeel Ahmed and Aisverya Soundararajan
Copyright © 2021 Jenny Stanford Publishing Pte. Ltd.
ISBN 978-981-4877-17-6 (Hardcover), 978-1-003-05694-2 (eBook)
www.jennystanford.com

for their physiochemical and biological characteristics, which in turn is responsible for their diversified applications. They are used in various industries such as pharmaceutical industry, food industry, and other non-food industries. With further advancements in technologies, polysaccharides have been structurally modified and studied for their various functions. In addition to these, the formation of nanocomposite with these polymers has been studied extensively for various applications from biosensors to drug delivery. They also find their importance in environmental applications by taking up several xenobiotics. Therefore, in this chapter, various polysaccharides from nature have been discussed in detail along with their applications.

1.1 Introduction

Polysaccharides are carbohydrate molecules made up of monosaccharide units bonded together by glycosidic linkages with a molecular weight of hundreds to thousands of Daltons [112, 149]. Depending on the number of monosaccharide units, they have been classified as linear or branched polysaccharides. They also have several functional groups such as hydroxyl group, carboxylic acid group, and amino group [103]. The naturally existing polysaccharides exhibit distinct features in terms of molecular weight, monosaccharide composition, glycosidic linkage types and patterns, configuration, charging properties, chain shapes, degree of polymerization, degree of branching, etc. These diversified structural properties of polysaccharides determine their functional properties such as solubility, flow behavior, gelling potential, surface and interfacial properties [54, 70].

Polysaccharides are the integral component of plant biomass, and 90% of the total polysaccharide produced on earth is constituted by vegetables. Apart from vegetables, they are produced by bacteria, fungi, algae, and animals [39, 99]. Different types of polysaccharides include starch, glycogen, cellulose, pectin, chitin, hemicellulose, lignin, etc [35]. Since ancient times, polysaccharides have been used in several industrial applications such as pharmaceuticals, food and nutrition, and other non-food industries. In the food industry, it has been used as emulsifiers, stabilizer, thickening agent, and as packaging materials [90, 104]. In addition to these, they play a vital role in the pharmaceutical industry due to their several advantageous characteristics such as biocompatibility, biodegradability, and ability of chemical modification [151]. They have been widely applied as binders, drug-release modifiers, thickeners, stabilizers, disintegrants, suspending agents, gelling agents, and as biodhesives [53]. Because of these wide ranges of applications, the market of polysaccharides market is continuously increasing. Therefore, it is essential to know the different types of polysaccharides produced by the biological entities. This chapter deals with the different types of natural polysaccharides and their applications in various fields.

1.2 Types of Polysaccharides

The various types of natural polysaccharides produced from different organisms are discussed as follows:

1.2.1 Plant Polysaccharides

Polysaccharides are the main constituents of vegetable biomass, and around 90% of cellulose has been obtained from plants. The various types of plant-based polysaccharides are discussed as follows:

1.2.1.1 Cellulose

Cellulose is the most abundant polysaccharide produced from plants. It was first discovered by the French chemist Anselme Payen when studying different types of wood. He found a substance that is not starch but can be broken into basic units of glucose. He named this compound "cellulose" [32, 132, 192]. In 1870, cellulose was commercially produced as a thermoplastic polymer by the Hyatt Celluloid Manufacturing Company [83]. The structure of cellulose is given in Fig. 1.1.

Payen found that cellulose is made up of 44–45% carbon, 6–6.5% hydrogen, and the remaining was oxygen. Based on this data, the molecular formula of cellulose was designated as $C_6H_{10}O_5$ [133]. Later in 1920, Haworth depicted the chain-like structure of cellulose and Staudinger produced the final structure of cellulose [60, 61,

166]. Since then it has been understood that cellulose is made up of repeating units of D-anhydroglucopyranose units (AGU) linked by β-1,4-glycosidic linkages with two ends. One end has C4–OH, which is alcohol borne and is called the non-reducing end, whereas the other terminating end has C1–OH made up of aldehyde group, called the reducing end [15, 84]. Some cellulose molecules were also found to have additional carbonyl and carboxyl groups [120]. Cellulose exists as chains, which vary with the source, and its molecular size depends on the degree of polymerization (DP). For example, the DP of cellulose present in cotton litter is 6500, whereas the cellulose in naturally occurring vascular plants has a DP higher than 10,000 [28, 89].

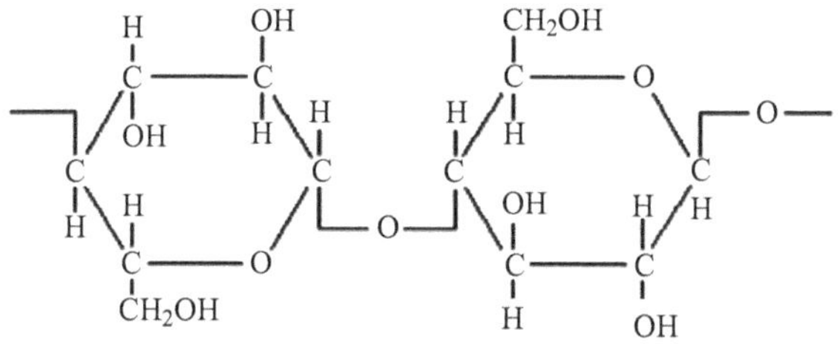

Figure 1.1 Structure of cellulose [141].

Cellulose is a hygroscopic material. It is insoluble in water but swells in water, acids, and in most solvents [145]. The poor solubility is attributed to strong intramolecular and intermolecular hydrogen bonding with individual chains [64]. The degradability of cellulose can be achieved by microbial enzymes. Partial degradation of cellulose by microbial enzymes enhances solubility of cellulose along with other advantageous properties [9]. Cellulose has better chirality, degradability, and chemical variability due to high reactivity from donor group [120]. Apart from these, they have other characteristics such as edibility, biodegradability, biocompatibility, and nontoxicity [2, 190]. These properties of cellulose make it useful in various food and pharmaceutical applications. Chemical modification of cellulose results in various cellulose derivatives such as carboxymethyl cellulose, microcrystalline cellulose, alkali cellulose, methyl cellulose, ethylmethylcellulose, hydroxypropylcellulose, sodium carboxymethylcellulose, etc. They have their own physical and chemical properties based on their modification [28].

Cellulose is obtained from several plant sources such as cotton, jute, ramie, flax, sisal, and hemp. It is also obtained from fique, agave, coconut coir, rice, barley, heat, straws, bamboo, grass, tree wood, and agriculture wastes [96]. Cellulose is abundantly utilized in the food and pharmaceutical industries. In the food industry, it has been used as stabilizers and emulsifiers to increase shelf life during storage. It is also used in white toppings, mousses, extruded foods, baked foods, bakery fillings, sauces, soups, egg white substitute, cereals, canned meats, beverages, frozen desserts, etc. [28]. In the pharmaceutical industry, it is used as fillers, binders, disintegrants, film-forming agent, thickeners, blocker, sustained release agent, blending agent, suspending agent, matrix and adhesives in tablets, and other formulations [55].

1.2.1.2 Hemicellulose

Hemicellulose is a part of wood fractions that consist of polysaccharide polymers with low degree of polymerization when compared with cellulose. It mainly consists of D-xylopyrannose, D-glucopyranose, D-galacopyranose, L-arabinofuranose, and D-mannopyranose. That is why they are called heterogenic group of polysaccharides [146, 182]. The non-cellulosic polysaccharides obtained from plant sources are grouped under hemicellulose. Hemicellulose represents around one third of the mass of cell walls and can be extracted easily with water and also alkali treatment, excluding glucan [88, 130, 175]. It can also be cleaved by action of enzymes such as xylan esterases, arabinofuranosidases, glucuronidases, and mannases [52, 162]. Hemicellulose is categorized into three types: mannan, xylan, and xyloglucan [131]. Apart from these, other polysaccharides such as galactans, arabinans, and arabinogalactans are also placed under hemicellulose [153].

The term hemicellulose has now been modified, and the term crosslinking glycans have been proposed instead of hemicellulose [181, 183]. The different structures of hemicellulose are given in Fig. 1.2. Xylose, mannose, and galactose form their main backbones, whereas arabinose, 4-O-methyl-D-glucuronic acid, and galctose form side chains [56].

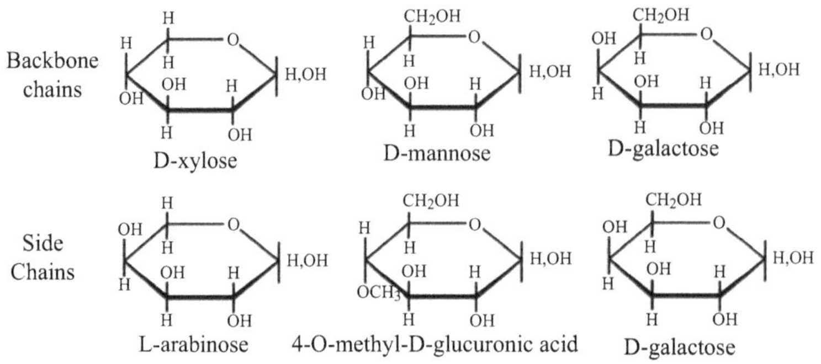

Figure 1.2 Structure of hemicellulose [56].

Mannan is the most ancient hemicellulose. It is abundant in almost all land plants and also in mosses and lycophytes [40, 57, 115]. Based on the structure of its backbone, it is classified as mannan, glucomannan, galactomannan, and galactoglucomannan [153]. It plays a vital role as a stabilizer and gelling agent in the food industry [17, 45]. Xylan is the major polysaccharide present in the cell walls of dicot plants and all types of walls in commelinid monocots [14]. Xylans are usually attached with arabinose with their backbone structure, and hence they are also called arabinoxylans or glucoarabinoxylans [44].

Xylans are used as additives to improve the flavor and baking performance of flour [31]. They are also used as probiotic and anti-carcinogenic agent in the pharmaceutical industry [46]. Apart from these, they are also used in several industrial applications due to their ability to form crosslinking films and gels [150]. Xyloglucon (XyG) is the main component of the primary cell wall in all dicot plants. It is also abundant in mosses and the primary walls of spermatophytes except grasses [48, 49, 115]. Xyloglucans are widely used in various kinds of food, pharmaceuticals, and industrial applications depending on the source of tissue and plant species [113]. In pharmaceuticals, they are used as carriers for drug-delivery systems, antibiotic administration, and also to treat ulcers [50, 114, 144].

1.2.1.3 Fructans

Fructans are sucrose-derived polysaccharides made up of repeating fructose units linked to a sucrose unit [142]. Fructose is present in most of the higher plants such as cereals (barley, oats, and wheat), vegetables (chicory, onion, and lettuce), ornamentals (dahlia and tulip), and forage grasses (*Lolium* and *Festuca*) [18, 63, 177]. Fructans get accumulated in various plants such as roots, stems, flowers, and seeds depending on the plant species and developmental stage [73, 105, 161, 179, 184, 187]. Fructans form polymers with various degrees of polymerization ranging from three to a few hundred depending on the species, developmental stage, and environmental conditions [178]. They are classified based on their glycosidic linkages as inulin, which has β (2,1), and levan, which has β (2,6), and a graminan type, which has both inulin and levan type of linkage [27, 76, 106, 173] (Fig. 1.3).

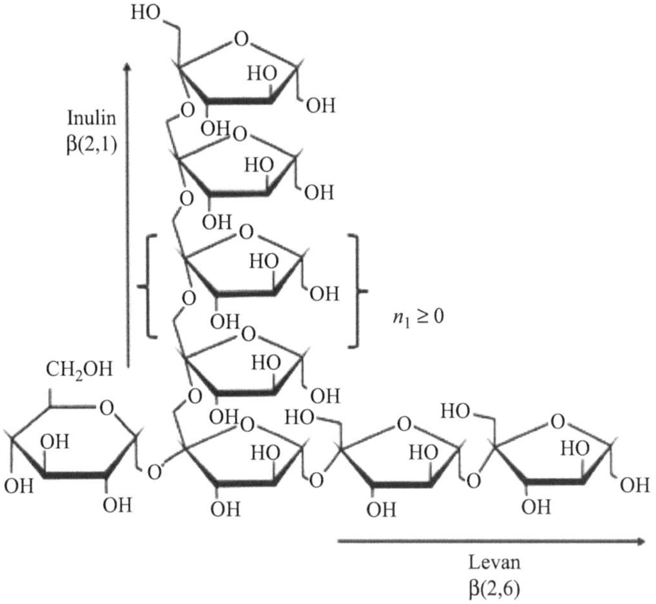

Figure 1.3 Structure of fructan with both levan and inulin linkages [173].

The physical and chemical properties of fructans vary with their degree of polymerization and the type of linkages they possess. The levan-type fructans are used as emulsifier and encapsulating agent in various products such as biodegradable plastics, cosmetics, glues, textile coatings, and detergents. They are also used in medical applications as blood plasma volume extender and also have antitumor and immunomodulatory activities in mice [19, 191]. Both levan and inulin are used as ideal substrates for producing high-fructose syrup. Inulin is produced in commercial scale as functional food with high nutritive value [143]. Fructans have also been proved as the best bifidogenic agent enabling them to be used as prebiotics [180].

1.2.1.4 Pectin

Pectin is a complex polysaccharide found in the middle lamella of cell wall. The concentration of pectin gradually decreases from primary wall to plasma membrane [81]. Pectins are produced in the initial stages of primary cell wall growth and form one third of the dry cell wall [72, 123]. Pectin is present in all dicot plants and also in monocots except grass membrane [81]. In the 1930s, the commercial production of pectin was started by Hermann Herbstreith who explained the use of pectin [20]. The structural prediction of pectin was difficult since it varies with source, during isolation, processing plant material, and storage [124, 171]. But pectins were thought to be made up of D-galacturonic acid joined together by α (1→4) linkage [10] (Fig. 1.4).

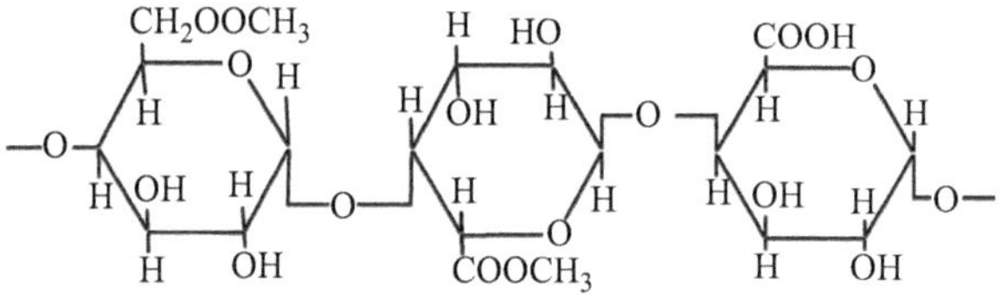

Figure 1.4 Structure of pectin [58].

Pectins are produced mainly from apple pomace and citrus peel [109]. But other sources such as banana, beet pulp, carrot, guava, lemon pulp, papaya, strawberries, sunflower rinds, tamarind, tomato fruit, etc. are also used for extraction of pectin [171]. Pectins from different sources may not have similar properties since the degree of methoxylation or esterification varies with different sources. Based on the degree of esterification, pectins are classified into two types: high methoxy pectins (HMPs), which have more than 50% carbonyl groups, and low methoxy pectins (LMPs), which have low carbonyl groups. The LMPs can further be converted to amidated pectin (AMP) with ammonia medium [167].

The application of pectins depends on the gel-forming ability [168]. The HMPs form gel immediately with sugar and acid, whereas the LMPs are unable to form gels since they do not possess sufficient acid groups to form gels [126]. They are used in preparing fruit gels, stabilizing acid emulsions such as salted cream and mayonnaise, increasing the texture and consistency of fruit products, etc. [38, 117]. They are not only used as a gelling agent but also as thickener, texturizer, emulsifier, and stabilizer in jams, jellies, fruit juices, frozen foods, confectionary products, beverages, barbecue sauce, etc. [65, 171]. In the pharmaceutical industry, they are used in removing lead and mercury from GIT [85], controlling hemorrhage [77], cholesterol reduction [163], gastric and duodenal ulcers [157], sustainable release of drugs, anti-obesity formulations, and personal care products [167].

1.2.2 Seaweed Polysaccharide

The study of marine organisms is called blue biotechnology. Marine organisms have been considered a valuable source for a diverse range of materials and have gained importance in food and medicine [129]. Since polysaccharides from marine origin are natural, nontoxic, biocompatible substance formed by several polymeric carbohydrate molecules linked together by glycosidic linkages, they are widely studied [122, 170]. The important polysaccharides from marine origin are listed as follows.

1.2.2.1 Alginate

Alginates were first described by British scientist E. C. C. Stanford in his patent dated 12 January 1881 [164, 165]. Since then several works on alginates have been performed, and it is considered the second most common exopolysaccharide produced by brown sea weeds of Class Pheaophyceae [66]. Alginates form the cell walls and intracellular matrix of brown sea weeds and provide mechanical strength and flexibility to the sea weeds [41]. Alginates are obtained from a wide range of algal sources such as *Laminaria digidata* [140], *Sargassum* spp. [95], *Undaria pinnatifida* [172], *Padina gymnospora* [188], *Dictyota dichotoms* [188], etc. Apart from these, they are also extracted from several bacterial community [147]. Alginate is a linear polymer consisting of two uronic acid units called D-mannuronic acid and L-glucuronic acid. These two units are linked by β(1→4) and α(1→4) glycosidic bonds [41]. The structure of alginate is provided in Fig. 1.5.

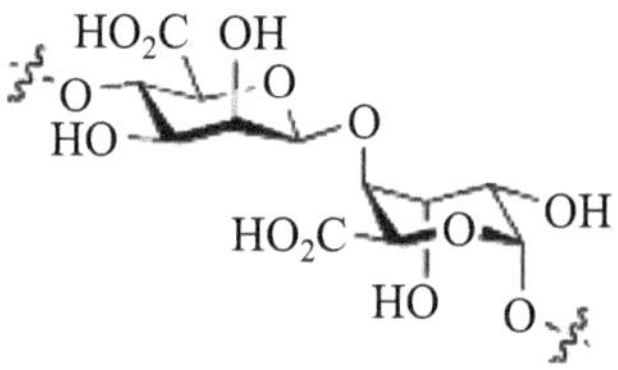

Figure 1.5 Structure of alginate [41].

They are located at the intracellular matrix of sea weeds as gel containing sodium, calcium, magnesium, strontium, and barium ions [59]. Alginates do not have any nutrient values, but due to their several advantageous properties such as solubility, adhesivity, biocompatibility, viscosity enhancement, gelling, and stabilizing ability, they were used to improve, modify, and stabilize the texture of some food products such as ice cream, sauces, ketchup, mayonnaise, frozen food, pastry fillings, puddings, pie fillings, etc. Apart from its application in the food industry, they also play a major role in the field of medicine as disintegrating agent, thickening agent, stabilizing agent, and antacid stomach protector in tablets, suspensions, emulsions, and capsules [111]. In modern medicine, they have been used as alginate matrices, microspheres, and microencapsulation for various processes such as drug delivery, protein delivery, tissue regeneration, tissue engineering, wound healing, wound dressings, etc. [4].

1.2.2.2 Carrageenans

Carrageenans are a type of sulfated polyglycans extracted from the species of red algae of the family Rhodophyceae. They are commonly found in Atlantic Ocean near Europe, Britain, and North America [176]. They have an ester sulfate content of 15–40% and an average molecular weight of more than 100 KDa (Fig. 1.6). Carrageenans are polymers of galactose with sulfate groups attached to its galactose backbone. Carrageenans are classified into various types based on their solubility in potassium chloride as λ, κ, ι, ε, and μ [8]. The primary difference in these types is based on the number and position of ester sulfate groups present as well as the content of 3,6-anhydro-D-galactose. In kappa carraggenan, D-galactose-4-sulfate and 3,6-anhydro-D-galactose are found as the major alternative sugar units, whereas iota carrageenan consists of a similar structure to kappa carrageenan except the 3,6-anhydro-D-galactose is attached to carbon 2 in the iota type. In lambda carrageenan, D-galactose-2-sulfate (a-1,3-linked) and D-galactose-2,6-disulfate (b-1,4-linked) exist as their primary structures [13].

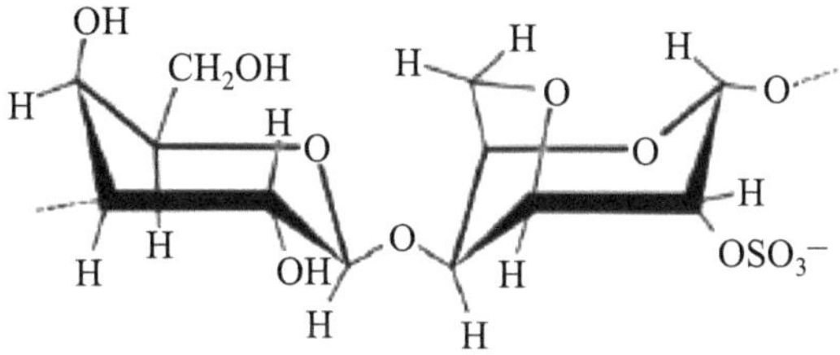

Figure 1.6 Structure of carrageenan [121].

Carrageenans have several advantageous properties such as better solubility, viscosity, gel-forming ability, etc. These properties of carrageenans made them to be used as texturizer, thickening agent, emulsifier, and stabilizer in various food products such as milk formulations, processed meat, etc. Apart from these they are also used in medicine as diabetic formulations, laxatives, and infant formula. In addition to these, they are also used in toothpaste, cosmetics, skin preparations, and pesticides [121, 189].

1.2.2.3 Fucans

Sulfated fucans, usually referred to as fucans or fucoidans, were initially named "fucoidin" and were first isolated by Killing in 1913. Now they have been named "fucoidan" according to the IUPAC system, but still other names such as fucans, sulfated fucans, and fucosan are also used [100]. Fucans are mainly extracted from the extracellular matrix of brown seaweed, jelly coat of sea urchin eggs, and body wall of sea cucumber [118, 135]. Fucan is a heterogenous macromolecule mainly made up of L-fucopyranosyl compounds linked covalently with sulfate molecules [135] along with other molecules such as galactose, mannose, and xylose (Fig. 1.7). The glucose molecules in fucans vary with different sources from which they have been isolated, place of isolation, and the methods used for isolating them [11, 128]. Not only the glucose molecules, the fucans isolated from different species also vary in their glycosidic pattern, sulfate content, site of sulfation, uronic acid content, and branching levels [12, 30, 134].

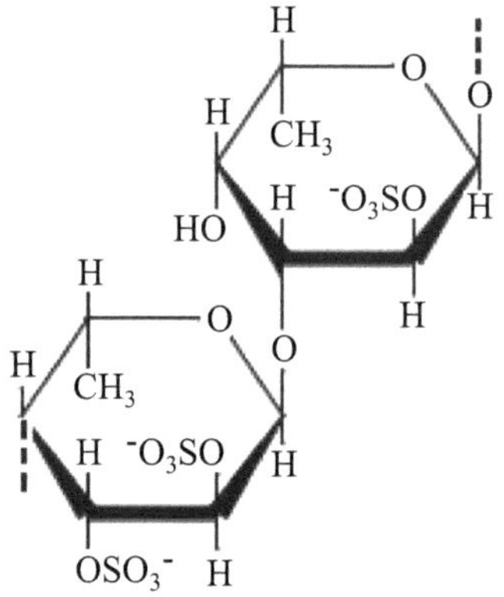

Figure 1.7 Fucoidan fraction from *A. nodosum* [11].

Fucans have several advantageous chemical properties such as water solubility, biocompatibility, and nontoxicity [21, 102]. The biological properties of fucans depend on their molecular weight, especially the number of sulfate groups present [25, 82]. They have several medicinal applications such as anticoagulant [186], antiviral [110], antitumor [29], antiadhesive [119], anti-inflammatory, antiproliferative [185], antioxidant [154], gastric ulcer protective, renal failure protective agent, and also have immunomodulatory activities [69, 194]. They are also used in food preparations to achieve better food preservation and prevent oxidation of stored food products [101].

1.2.2.4 Ulvans

The name "ulvan" was originally coined as "ulvin" or "ulvacin" by Kylin (1946) referring to the different fractions of sulfated polysaccharide extracted from the cell walls of *Ulva lactuca* [94]. Ulvans are mainly extracted from two genera of the family Ulva and Enteromorpha. This complex hydrocolloid was found to consist of glucoronoxylrhamnon units along with cellulose (Fig. 1.8). Ulvans are sulfated polysaccharides consisting of glucose, iduronic acid, and sulfated rhamnose units [93, 94]. The presence of iduronic acid makes ulvan a unique polysaccharide compared to the other polysaccharides of marine origin. The average molecular weight of ulvans was found to range between 189 and 8900 KDa, and the cell wall of Ulvales was found to contain around 38–54% of ulvans as dry algal matter [22, 158]. The molecular weight and composition of ulvan vary with electron density, charge distribution, taxonomy, growth conditions, period of collection, and post-collection treatment [24, 33, 42, 43].

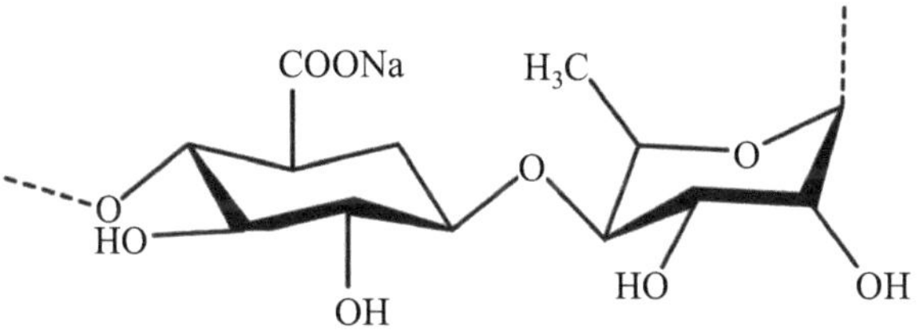

Figure 1.8 Structure of ulvan [155].

Since their discovery, ulvans have been studied for their composition, methods of extraction, and biological activities. It has been found that the biological activities of ulvans play a vital role in the pharmaceutical industry since ulvan exhibits antioxidant activity, anticoagulant activity, antihyperlipidemic activity, antiviral activity, antinociceptive, antiprotozoal, antibacterial, antitumor activity, immune stimulation, tissue engineering, wound dressing, best gelling ability, etc. [3, 34, 174]. Since ulvan is the primary intracellular energy storage for algal cells, it is used in biogas and bioenergy production. In addition, ulvans are also used as supplements for human and animal consumption [16, 108].

1.2.3 Microbial Polysaccharides

A wide range of exopolysaccharides are produced as extracellular or cell surface material by the microbial community. These microbial polysaccharides are gaining importance rather than the polysaccharides of other sources due to their unique and superior physical properties along with easy, economic, and controlled production [116]. Like other polysaccharides, microbial exopolysaccharides have also been used in a wide range of applications in food and non-food industries. Some important microbial polysaccharides are discussed as follows.

1.2.3.1 Pullulan

Pullulan is one of the commercially emerging biopolymers produced by the fungal species *Aureobasidium pullulans.* Pullulan (Fig. 1.9) is a water-soluble neutral glucan consisting of either linear chains of glucopyranose units with regular alteration of two α (1 → 4) and one (1 → 6) linkages or a linear polymer of maltotriose units connected by (1 → 6) linkages [193]. Apart from *A. pullulans*, it is also produced from other species such as *Tremella mesenterica* [47], *Cryphonectria parasitica* [37], *Teloschites flavians* [139], etc.

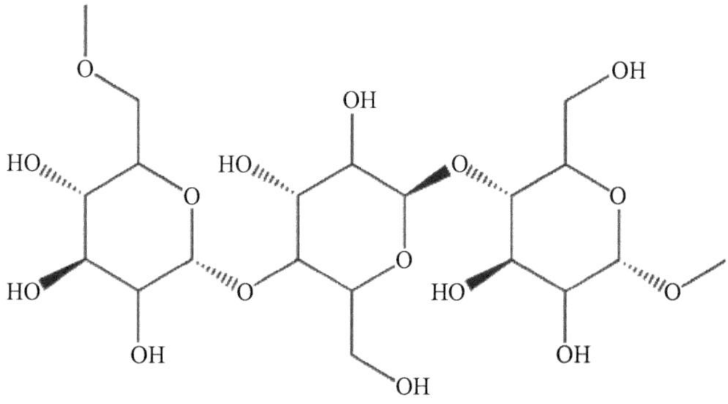

Figure 1.9 Structure of pullulan [67].

Pullulans are readily soluble in water and have better viscosity. They are also biodegradable, oxygen impermeable, and edible. For these properties, pullulans are used in various food applications such as non-polluting wrapping material for food supplements, binder, lubricant, gelling agent, emulsifier, and stabilizer for various food products [98, 159]. In the pharmaceutical industry, pullulan is used as denture adhesive, oral care strips, drug delivery, gene delivery, vaccination, medical imaging, molecular chaperones, and film forming [23, 78, 86, 87, 152, 156, 160].

1.2.3.2 Xanthan gum

Xanthan gum is an extracellular heteropolysaccharide produced by *Xanthomonas campestris*. Xanthan gum is made up of repeating subunits of 3 to 8 monosaccharides [97, 169]. The primary structure of xanthan gum (Fig. 1.10) consists of linear β (1→4) linked D-glucose backbone with trisacchairde chain on every glucose at C3-containing glucuronic acid residue linked to terminal mannose units [91, 148]. Xanthan gums are highly viscous even at low concentrations, are highly soluble in water, and have better stability and defrosting. For these abilities, xanthan gums are exploited in the food industry as thickeners, emulsifier, and stabilizer in ketchups, dairy products, beverages, chocolates, desserts, jellies, margarine, yoghurt, bakery products, frozen foods, sauces, and gravies. Blends of xanthan gum with other polysaccharides act as an excellent stabilizer for ice cream, ice milk, sherbet, and milk shakes.

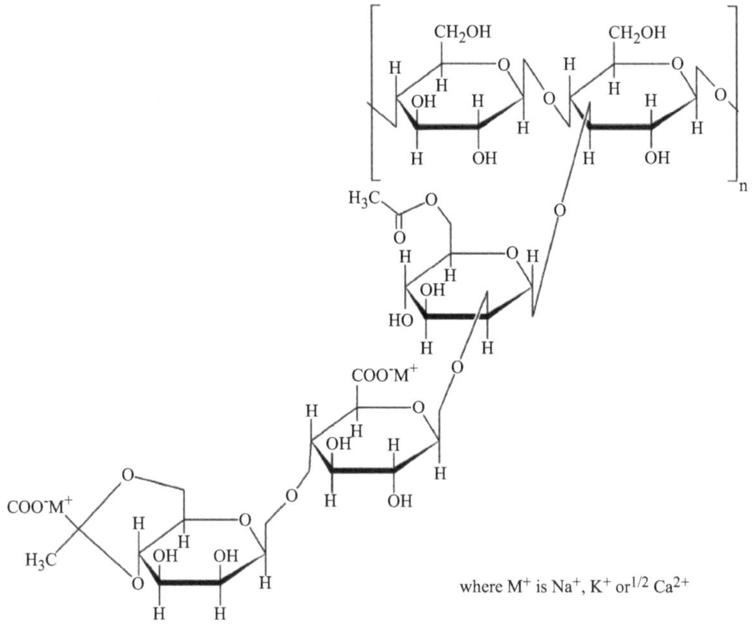

Figure 1.10 Structure of xanthan gum [138].

Xanthan, along with guar and LBG blend, is vital to slice ability, firm body, and flavor release of cream cheese. Also, xanthan thickens cottage cheese dressings by providing good drainage control [5, 26, 36, 107, 127]. Apart from these, it is also used in pharmaceutical and biomedical applications that involve digestion [1].

1.2.3.3 Gellan gum

Gellan is an exopolysaccharide discovered in 1978 from the bacterium *Sphingmonas paucimobilis* [137]. It is made up of β-D-glucose, L-rhamnose, and D-glucuronic acid units with an approximate composition of glucose (60%), rhamnose (20%), and glucuronic acid (20%). It also contains acyl, acetate, and glycerate groups bound to the glucose residue adjacent to glucuronic acid [71, 125]. The above-mentioned composition is seen only in natural gellan gums (Fig. 1.11). Gellan gums can be classified into further two types as deacetylated gellan gum in which acyl groups are removed from their structure and clarified gellan gums, which results from hot treatment of deacetylated gellan gums [79].

(a)

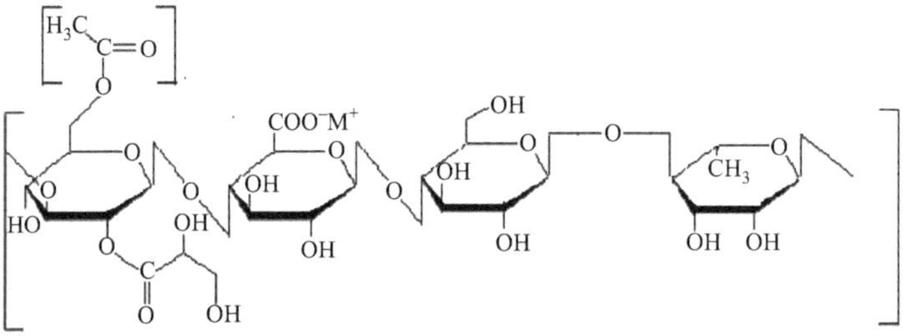

(b)

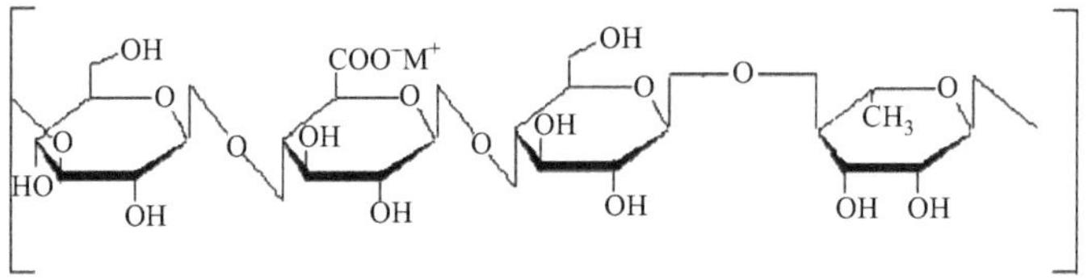

Figure 1.11 (a) Structure of gellan gum and (b) structure of deacetylated gellan gum [136].

Gellan gums are used in various food products such as confectionaries, jams, jellies, fabricated foods, dairy products, icing and toppings, pet foods, and also in the reduction of oil uptake while frying. In the pharmaceutical industry, they are used for sustained drug release, ophthalmic formulations, and as binding agents in tablet formulations. Apart from these, they are also utilized in immobilization and microencapsulation techniques. They are also used as an alternative to agar in the preparation of media [7, 51].

1.2.4 Animal Polysaccharide

Polysaccharides from animal origin, i.e., shell wastes of marine organisms such as arthropods and crustaceans have been widely studied. It is notable that like other polysaccharides, those from shell wastes also produce various advantageous characteristics and they are widely used in food and pharmaceutical applications. The important animal polysaccharides are considered to be chitin and chitosan, which are discussed as follows.

1.2.4.1 Chitin and chitosan

Chitin and chitosan are important polysaccharides present in arthropods and crustaceans (Fig. 1.12). Though they are similar in structure to cellulose, they are not produced by organisms producing cellulose [68]. Chitin was discovered by Henri Braconnot, a French professor, in 1811 after the discovery of a "material particularly resistant to usual chemicals" by A. Hachett, an English scientist, in 1799. Chitosan was first discovered by Rouget in 1859 when he was experimenting with the chemical and thermal manipulation of chitin [6, 74]. Chitin is a natural biopolysaccharide consisting of β-(1–4)-poly-N-acetyl-D-glucosamine, whereas chitosan is derived by the deacetylation of chitin polysaccharide [92]. The degree of deacetylation is determined by the proportion of D-glucosamine and N-acetyl D-glucosamine [75].

Since the introduction of chitin and chitosan, it has been explored for several applications. The ability of chitosan to bind with a variety of substances, including acids, lipophilic substances, minerals, etc., and its physical, chemical properties have made chitosan to be used in several fields such as food, medicine, cosmetics, agriculture, paper, and also fabrics [75]. In recent years, several derivatives of chitin and chitosan have been used for gene and drug delivery [62].

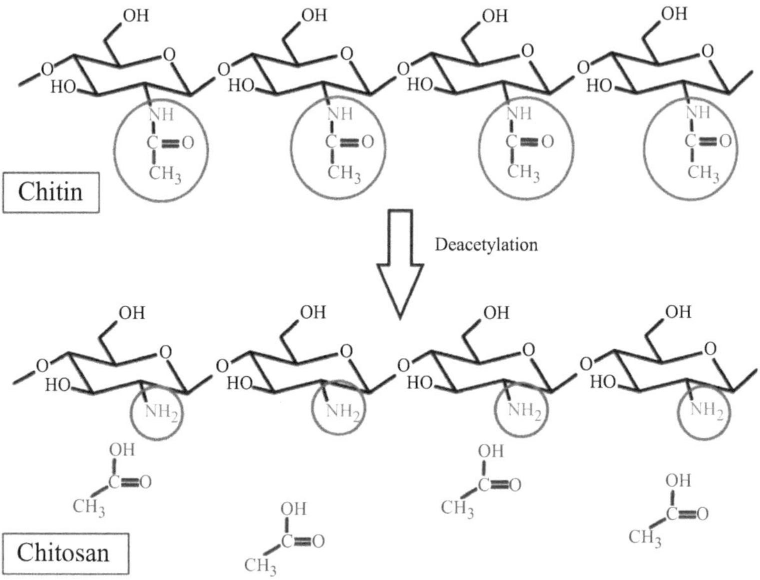

Figure 1.12 Structure of chitin and chitosan [80].

1.3 Conclusion

Polysaccharides are found to have various properties such as stability, gelling ability, viscosity, biodegradability, and biocompatibility. Despite their origin of sources, they can be utilized in various nutritional and pharmaceutical applications. With further advancements, they can be produced in higher quantities and can be explored for more applications in various fields of science.

References

1. Ahmad, N. H., Mustafa, S., and Man, Y. B. C. (2013). Microbial polysaccharides and their modification approaches: A review. *Int. J. Food Prop.*, DOI: 10.1080/10942912.2012.693561.
2. Albuquerque, P. B. S., Coelho, L. C. B. B., Teixeira, J. A., and Carneiro-da-Cunha, M. G. (2016). Approaches in biotechnological applications of natural polymers. *AIMS Mol. Sci.*, **3**, 386–425.
3. Alves, A., Pinho, E. D., Neves, N. M., Sousa, R. A., and Reis, R. L. (2012). Processing ulvan into 2D structures: Cross-linked ulvan membranes

as new biomaterials for drug delivery applications. *Int. J. Pharm.*, **426**, 76–81.

4. Ashwini, R., Vijayanand, S., Sudha, P. N., Aisverya, S., and Hemapriya, J. (2019). Neutraceutical efficiency of fucan polysaccharides from marine sources, In: *Marine Polysaccharides: Advances and Multifaceted Applications* (Ahmed, S. and Soundararajan, A., Eds.), Jenny Stanford Publishing, Singapore.
5. Association of Official Analytical Chemists. (2007). *Official Methods of Analysis*, 18th edition, Chapter 33, Benjamin Franklin Station, Washington, D.C., USA, 75, 79, 82, 85.
6. Au Natural Herbals. (2001). Chitosan history. Available from URL: http://www.chitosan-weightloss.net/history.html.
7. Bajaj, I. B., Survase, S. A., Saudagar, P. S., and Singhal, R. S. (2007). Gellan gum: Fermentative production, downstream processing and applications. *Food Technol. Biotechnol.*, **45**(4), 341–354.
8. Barbeyron, T., Michel, G., Potin, P., Henrissat, B., and Kloareg, B. (2000). ι-Carrageenases constitute a novel family of glycoside hydrolases, unrelated to that of κ-carrageenases. *J. Biol. Chem.*, **275**, 35499–35505.
9. Becker, D., Braet, C., Brumer III, H., Clacyssens, M., Divne, C., Fagerstrom, B. R., Harris, M., Jones, T. A., Kleywegt, G. J., Koivula, A., Mahdi, S., Piens, K., Simott, J. Stahlberg, M. L., Teeri, T. T., Underwood, M., and Wohlfahrt, G. (2006). Engineering of a glycosidose family 7 cellobiohydrolose, to more alkaline pH optimum: The pH behaviour of *Trichoderma reesei* Cel7A and its mutant. *Biochem. J.*, **356**, 19.
10. Belitz, H. D. W. G. and Schieberle, P. (2004). Chemistry of pectin. In: *Food Chemistry*, 4th revised extended edition, Springer, Berlin.
11. Berteau, O. and Mulloy, B. (2003). Sulfated fucans, fresh perspectives: Structures, functions and biological properties of sulphated fucans and overview of enzymes active towards this class of polysaccharide. *Glycobiology*, **13**(6), 29–40.
12. Bilan, M. I. and Usov, A. I. (2008). Structural analysis of fucoidans. *Nat. Prod. Commun.*, **3**, 1639–1648.
13. Blakemore, W. R. and Harpell, A. R. (2010). Carrageenan. In: *Food Stabilisers, Thickeners and Gelling Agent* (Imeson, A., Ed.), Wiley-Blackwell, Blackwell Publishing Ltd., Published Online, 73–94.
14. Brown, D. M., Goubet, F., Wong, V. W., Goodacre, R., Stephens, E., Dupree, P., and Turner, S. R. (2007). Comparison of five xylan synthesis mutants reveals new insight into the mechanisms of xylan synthesis. *Plant J.*, **52**, 1154–1168.

15. Brown, G. and Ley, H. (1965). α-D-glucose: Precise determination of crystal and molecular structure by neutron diffraction analysis. *Science*, **147**, 1038–1039.

16. Bruhn, A., Dahl, J., Nielsen, H. B., Nikolaisen, L., Rasmussen, M. B., Markager, S., Olesen, B., Arias, C., and Jensen, P. D. (2011). Bioenergy potential of *Ulva lactuca*, biomass, yield, methane production and combustion. *Biores. Technol.*, **102**, 2595–2604.

17. Buckeridge, M. S. (2010). Seed cell wall storage polysaccharides: Models to understand cell wall biosynthesis and degradation. *Plant Physiol.*, **154**, 1017–1023.

18. Cairns, A. J., Pollock, C. J., Gallagher, J. A., and Harrison, J. (2000). Fructans: Synthesis and regulation. In: *Photosynthesis: Physiology and Metabolism* (Leegood, R. C., Sharkey, T. D., and von Caemmerer, S.), Kluwer Academic Publishers, 301–320.

19. Calazans, G. M. T., Lima, R. C., De Franca, F. P., and Lopes, C. E. (2000). Molecular weight and antitumour activity of *Zymomonas mobilis* levans. *Int. J. Biol. Macromol.*, **27**, 245–247.

20. Canteri, M. H. G., Moreno, L. R., Wosiacki, G., and Scheer, A. P. (2012). Pectin: From raw material to the final product. *Polímeros*, **22**(2), 1–13.

21. Cao, R. A., Lee, Y. J., and You, S. G. (2014). Water soluble sulfated-fucans with immune-enhancing properties from Ecklonia cava. *Int. J. Biol. Macromol.*, **67**, 303–311.

22. Cardoso, M. J., Costa, R. R., and Mano, J. F. (2016). Marine origin polysaccharides in drug delivery systems. *Mar. Drugs*, **14**(2), 34.

23. Chatap, V. K., Maurya, A. R., Deshmukh, P. K., and Zawar, L. R. Formulation and evaluation of nisoldipne sublingual tablets using pullulan and chitosan for rapid oromucosal absorption. *Adv. Pharmacol. Pharm.*, **1**, 18–25.

24. Chiellinim, F. and Morelli, A. (2011). Ulvan: A versatile platform of biomaterials from renewable resources. In: *Biomaterials: Physics and Chemistry* (Pignatello, R., Ed.), IntechOpen, DOI: 10.5772/24901. Available from: https://www.intechopen.com/books/biomaterials-physics-and-chemistry/ulvan-a-versatile-platform-of-biomaterials-from-renewable-resources.

25. Cho, M., Han, J. H., and You, S. (2011). Inhibitory effects of fucan sulphates on enzymatic hydrolysis of starch. *LWT—Food Sci. Technol.*, **44**, 1164–1171.

26. Silva, M. F., Fornari, R. C. G., Mazutti, M. A., de Oliveira, D., Padilha, F. F., Cichoski, A. J., Cansian, R. L., Di Luccio, M., and Treichel, H. (2009).

Production and characterization of xanthan gum by *Xanthomonas campestris* using cheese whey as sole carbon source. *J. Food Eng.*, **90**(1), 119–123.

27. Cochrane, M. P. (2000). Seed carbohydrates. In: *Seed Technology and Its Biological Basis* (Black, M. and Bewley, J. D. Eds.), Boca Raton, FL: CRC Press, 85–120.
28. Coffey, D. G., Bell, D. A., and Henderson, A. (2006). Cellulose and cellulose derivatives, In: *Food Polysaccharides and Their Applications*, 2nd edition, CRC Press, 147–174.
29. Coombe, D. R., Parish, C. R., Ramshaw, I. A., and Snowden, J. M. (1987). Analysis of the inhibition of tumor metastasis by sulfated polysaccharides. *Int. J. Cancer*, **39**, 82–88.
30. Costa, L. S., Fidelis, G. P., Cordeiro, S. L., Oliveira, R. M., Sabry, D. A., Camara, R. B. G., Nobre, L. T. D. B., Costa, M. S. S. P., Almeida-Lima, J., Farias, E. H. C., Leite, E. L., and Rocha, H. A. O. (2010). Biological activities of sulfated polysaccharides from tropical seaweeds. *Biomed. Pharma.*, **64**, 21–28.
31. Courtin, C. M. and Delcour, J. A. (2002). Arabinoxylans and endoxylanases in wheat flour bread-making. *J. Cereal Sci.*, **35**, 225–243.
32. Crawford, R. L. (1981). *Lignin Biodegradation and Transformation*, John Wiley and Sons, New York.
33. Cunha, L. and Grenha, A. (2016). Sulfated seaweed polysaccharides as multifunctional materials in drug delivery applications. *Mar. Drugs*, **14**(3), DOI: 10.3390/md14030042.
34. Dash, M., Samal, S. K., Bartoli, C., Morelli, A., Smet, P. F., Dubruel, P., and Chiellini, F. (2014). Biofunctionalization of ulvan scaffolds for bone tissue engineering. *ACS Appl. Mater. Interfaces*, **6**, 3211–3218.
35. Dave, D. T. and Shah, G. B. (2015). Pharmacological potential of naturally occurring non-starch polysaccharides (NSP). *J. Phytopharmacol.*, **4**(6), 307–310.
36. Deckwer, W. D. (1989). The influence of agitation rate on xanthan production by *Xanthomonas campestris*. *Biotechnol. Bioeng.*, **34**(11), 1393–1397.
37. Delben, F., Forabosco, A., Guerrini, M., Liut, G., and Torri, G. (2006). Pullulans produced by strains of *Cryphonectria parasitica*-II. Nuclear magnetic resonance evidence. *Carbohyd. Polym.*, **63**, 545–554.
38. Dhame, A. (1992). Gelpoint measurements on high methoxyl pectin by different techniques. *J. Text. Stud.*, **23**, 1.

39. Di Donato, P., Poli, A., Taurisano, V., and Nicolaus, B. (2014). Polysaccharides: Applications in biology and biotechnology/ polysaccharides from bioagro-waste new biomolecules-life, *Polysaccharides*, DOI: 10.1007/978-3-319-03751-6_16-1.

40. Domozych, D. S., Ciancia, M., Fangel, J. U., Mikkelsen, M. D., Ulvskov, P., and Willats, W. G. (2012). The cell walls of green algae: A journey through evolution and diversity. *Front Plant Sci.*, **3**, 82.

41. Draget, K. I., Smidsrød, O., and Skjåk-Bræk, G. (2005). Alginates from algae. In: *Polysaccharides and Polyamides in the Food Industry. Properties, Production, and Patents* (Steinbüchel, A. and Rhee, S. K., Eds.), Wiley, Verlag.

42. Dubigeon, B. C., Lahaye, M., and Barry, J. L. (1997). Human colonic bacterial degradation of dietary fibres from sea-lettuce (*Ulva* sp.). *J. Sci. Food Agric.*, **73**, 149–159.

43. Dubigeon, B. C., Lahaye, M., Guillon, F., Barry, J. L., and Gallant, D. J. (1997). Factors limiting the biodegradation of *Ulva* sp. cell wall polysaccharides. *J. Sci. Food Agric.*, **75**, 341–351.

44. Ebringerova, A., Hromadkova, Z., and Heinze, T. (2005). Hemicellulose. *Adv. Polym. Sci.*, **186**, 1–67.

45. Edwards, M., Scott, C., Gidley, M. J., and Reid, J. S. G. (1992). Control of mannose galactose ratio during galactomannan formation in developing legume seeds. *Planta*, **187**, 67–74.

46. Fooks, L. J., Fuller, R., and Gibson, G. R. (1999). Prebiotics, probiotics and human gut microbiology. *Int. Dairy J.*, **9**, 53–61.

47. Fraser, C. G. and Jennings, H. J. (1971). A glucan from *Tremella mesenterica* NRRL-Y6158. *Can. J. Chem.*, **49**, 1804–1807.

48. Fry, S. C. and Janice, G. M. (1989). Toward a working model of the growing plant cell wall. In: *Plant Cell Wall Polymers*, American Chemical Society, Washington, D.C., 33–46.

49. Fry, S. C., York, W. S., Albersheim, P., Darvill, A., Hayashi, T., Joseleau, J.-P., Kato, Y., Lorences, E. P., Maclachlan, G. A., McNeil, M., Mort, A. J., Reid, J. S. G., Seitz, H. U., Selvendran, R. R., Voragen, A. G. J., and White, A. R. (1993). An unambiguous nomenclature for xyloglucan-derived oligosaccharides. *Physiol. Plant.*, **89**, 1–3.

50. Ghelardi, E., Tavanti, A., Celandroni, F., Lupetti, A., Blandizzi, C., Boldrini, E., Campa, M., and Senesi, S. (2000). Effect of a novel mucoadhesive polysaccharide obtained from tamarind seeds on the intraocular penetration of gentamicin and ofloxacin in rabbits. *J. Antimicrob. Chemother.*, **46**, 831–834.

51. Giavasis, L., Harvey, L. M., and McNeil, B. (2000). Gellan gum. *Crit. Rev. Biotechnol.*, **20**(3), 177–211.

52. Goldbeck, R., Damásio, A. R. L., Gonçalves, T. A., Machado, C. B., Paixão, D. A. A., Wolf, L. D., Mandelli, F., Rocha, G. J. M., Ruller, R., and Squina, F. M. (2014). Development of hemicellulolytic enzyme mixtures for plant biomass deconstruction on target biotechnological applications. *Appl. Microbiol. Biotechnol.*, **98**(20), 8513–8525.

53. Guo, J., Skinner, G. W., Harcum, W. W., and Barnum, P. E. (1998). Pharmaceutical applications of naturally occurring water-soluble polymers. *Pharm. Sci. Technol. Today*, **1**, 254–261.

54. Guo, M. Q., Hu, X., Wang, C., and Ai, L. (2017). Polysaccharides: Structure and solubility. In: *Solubility of Polysaccharides*, IntechOpen, 7–21. http://dx.doi.org/10.5772/intechopen.71570.

55. Gupta, P. K., Raghunath, S. S., Prasanna, D. V., Venkat, P., Shree, V., Chithananthan, C., Choudhary, S., Surender, K., and Geetha, K. (2019). An update on overview of cellulose, its structure and applications. In: *Cellulose*, IntechOpen, 1–22.

56. Gurunathan, T., Mohanty, S., and Nayak, S. K. (2015). A review of the recent developments in biocomposites based on natural fibres and their application perspectives. *Compos. Pt. A Appl. Sci. Manuf.*, **77**, 1–25.

57. Harholt, J., Sørensen, I., Fangel, J., Roberts, A., Willats, W. G., Scheller, H. V., Petersen, B. L., Banks, J. A. and Ulvskov, P. (2012). The glycosyltransferase repertoire of the spikemoss Selaginella moellendorffii and a comparative study of its cell wall. *PLoS One*, **7**(5), e35846.

58. Hassan, B., Chatha, S. A. S., Hussain, A. I., Zia, K. M., and Akhtar, N. (2018). Recent advances on polysaccharides, lipids and protein based edible films and coatings: A review. *Int. J. Biol. Macromol.*, **109**, 1095–1107.

59. Haug, A. (1964). Composition and Properties of Alginates, Thesis, Norwegian Institute of Technology, Trondheim.

60. Haworth, W. N. (1928). The structure of carbohydrates. *Helv. Chim. Acta*, **11**, 534–548.

61. Haworth, W. N. (1932). Die Konstitution einiger Kohlenhydrate. *Ber. Dtsch. Chem. Ges. (A)*, **65**, 43.

62. Hemapriya, J., Ravi, A., Sudha, P. N., Aisverya, S., and Vijayanand, S. (2019). Chitosan and its biomedical applications. In: *Marine*

Polysaccharides: Advances and Multifaceted Applications (Ahmed, S. and Soundararajan, A.), Jenny Stanford Publishing, Singapore.

63. Hendry, G. A. F. and Wallace, R. K. (1993). The origin, distribution, and evolutionary significance of fructans. In: *Science and Technology of Fructans* (Suzuki, M. and Chatterton, N. J.), CRC Press, Boca Raton, FL, 119–139.
64. Hinterstoisser, B. and Salmen, L. (2000). Application of dynamic 2D FTIR to cellulose. *Vib. Spectrosc.*, **22**, 111–118.
65. Hoefler, A. C. (1991). Other pectin food products. In: *The Chemistry and Technology of Pectin* (Walter, R. H., Ed.), Academic Press, San Diego, 51–66.
66. Holdt, S. L. and Kraan, S. (2011). Bioactive compounds in seaweed: Functional food applications and legislation. *J. Appl. Phycol.*, **23**(3), 543–597.
67. https://www.chemicalbook.com/ChemicalProductProperty_US_CB8141488.aspx
68. Hudson, S. M. and Smith, C. (1998). *Biopolymers from Renewable Resources* (Kaplan, D. L., Ed.), Springer–Verlag, 96–118.
69. Hwang, H., Kwon, M., Kim, I., and Nam, T. (2008). The effect of polysaccharide extracted from the marine alga *Capsosiphon fulvescens* on ethanol administration. *Food Chem. Toxicol.*, **46**, 2653–2657.
70. Izydorczyk, M., Cui, S. W., and Wang, Q. (2005). Polysaccharide gums: Structures. In: *Functional Properties and Applications*, Taylor and Francis.
71. Jansson, P. E., Lindberg, B., and Sandford, P. A. (1983). Structural studies of gellan gum, an extracellular polysaccharide elaborated by *Pseudomonas elodea*. *Carbohydr. Res.*, **124**, 135–139.
72. Jarvis, M. C., Forsyth, W., and Duncan, H. J. (1988). A survey of the pectin content of nonlignified monocot cell walls. *Plant Physiol.*, **88**, 309–341.
73. Jenkins, C. L., Lewis, D., Bushell, R., Belobrajdic, D. P., and Bird, A. R. (2011). Chain length of cereal fructans isolated from wheat stem and barley grain modulates *in vitro* fermentation. *J. Cereal Sci.*, **53**(2), 188–191.
74. Jeuniaux, C. (1996). A brief survey of the early contribution of European scientists to chitin knowledge. In: *Advances in Chitin Sciences* (Domard, A., Jeuniaux, C., Muzzarelli, R. A. A., Roberts, G., Eds.), Jacques André Publ., Lyon, France, 1–9.

75. Jing, S., Li, L., Ji, D., Takiguchi, Y., and Yamaguchi, T. (1997). Effect of chitosan on renal function in patients with chronic renal failure. *J. Pharm. Pharmacol.*, **49**, 721–723.

76. John, P. (1992). *Biosynthesis of the Major Crop Products*. Wiley, Chichester, UK, 114–126.

77. Joseph, G. H. (1956). *Pectin Bibliography of Pharmaceutical Literature*, Sunskist Growers, Ontario, CA.

78. Kageyama, S., Wada, H., Muro, K., Niwa, Y., Ueda, S., Miyata, H., Takiguchi, S., Sugino, S. H., Miyahara, Y., Ikeda, H., Imai, N., Sato, E., Yamada, T., Osako, M., Ohnishi, M., Harada, N., Hishida, T., Doki, Y., and Shiku, H. (2013). Dose-dependent effects of NY-ESO-1 protein vaccine complexed with cholesteryl pullulan (CHP-NY-ESO-1) on immune responses and survival benefits of esophageal cancer patients. *J. Transl. Med.*, **11**, 246.

79. Kang, K. S., Veeder, G. T., Mirrasoul, P. J., Kaneko, T., and Cottrell, L. W. (1982). Agar-like polysaccharides produced by a *Pseudomonas* species: Production and basic properties. *Appl. Environ. Microbiol.*, **43**, 1086–1091.

80. Kaur, S. and Dhillon, G. S. (2013). Recent trends in biological extraction of chitin from marine shell wastes: A review. *Crit. Rev. Biotechnol.*, DOI: 10.3109/07388551.2013.798256.

81. Kertesz, Z. I. (1951). The pectic substances, Interscience, New York.

82. Kim, T. H. and Bae, J. S. (2010). Ecklonia cava extracts inhibit lipopolysaccharide induced inflammatory responses in human endothelial cells. *Food Chem. Toxicol.*, **48**, 1682–1687.

83. Klemm, D., Heublein, B., Fink, H. P., and Bohn, A. (2005). Cellulose: Fascinating biopolymer and sustainable raw material. *Chem. Inform.*, **36**(36), DOI: 10.1002/chin.200536238.

84. Klemm, D., Philipp, B., Heinze, T., Heinze, U., and Wagenknecht, W. (1998). *Comprehensive Cellulose Chemistry, Volume 1, Fundamentals and Analytical Methods*, Wiley-VCH, Weinheim.

85. Kohn, R. (1982). Binding of toxic cations to pectin, its oligomeric fragments and plant tissues, *Carbohydr. Polym.*, **2**, 273–275.

86. Kong, I. G., Sato, A., Yuki, Y., Nochi, T., Takahashi, H., Sawada, S., Mejima, M., Kurokawa, S., Okada, K., Sato, S., Briles, D. E., Kunisawa, J., Inoue, Y., Yamamoto, M., Akiyoshi, K., and Kiyono, H. (2013). Nanogel-based PspA intranasal vaccine prevents invasive disease and nasal colonization by *Streptococcus pneumonia*. *Infect. Immun.*, **81**, 1625–1634.

87. Kong, S. H., Noh, Y. W., Suh, Y. S., Park, H. S., Lee, H. J., Kang, K. W., Kim, H. C., Lim, Y. T., and Yang, H. K. (2015). Evaluation of the novel near-infrared fluorescence tracers pullulan polymer nanogel and indocyanine green/γ-glutamic acid complex for sentinel lymph node navigation surgery in large animal models. *Gastric Cancer*, **18**, 55–64.

88. Kooiman, P. (1961). The constitution of Tamarindus-amyloid. *Recl. Trav. Chim. Pay. B*, **80**, 849–865.

89. Krässig, H. (1996). *Cellulose, Polymer Monographs*, Vol. 11, Gordon and Breach Science Publishers, Amsterdam, 6–42.

90. Kroll, J. (1991). Microemulsions and emulsions in foods. In: *ACS Symposium Series*, Vol. 448 (El-Nokaly, M. and Cornell D., Eds.), American Chemical Society, Washington.

91. Kumar, A. S. and Mody, K. (2009). *Microbial Production of Biopolymers and Polymer Precursors: Applications and Perspectives*, Caister Academic Press, Norfolk, UK, 229–254.

92. Kumar, M. N. V. R., Muzzarelli, R. A. A., Muzzarelli, C., Sashiwa, H., and Domb, A. J. (2004). Chitosan chemistry and pharmaceutical perspectives. *Chem. Rev.*, **104**, 6017–6084.

93. Lahaye, M. (1998). NMR spectroscopic characterisation of oligosaccharides from two *Ulva rigida* ulvan samples (Ulvales, Chlorophyta) degraded by a lyase. *Carbohydr. Res.*, **314**, 1–12.

94. Lahaye, M. and Robic, A. (2007). Structure and functional properties of ulvan, a polysaccharide from green seaweeds. *Biomacromolecules,* **8**, 1765–1774.

95. Latifi, A. M., Nejad, E. S., and Babavalian, H. (2015). Comparison of extraction different methods of sodium alginate from brown alga *Sargassum* sp. localized in the Southern of Iran. *J. Appl. Biotechnol. Rep.*, **2**(2), 251–255.

96. Lavanya, D., Kulkarni, P. K., Dixit, M., Raavi, P. K., and Krishna, L. N. V. (2011). Sources of cellulose and their applications: A review. *IJDFR*, **2**(6), 19–38.

97. Laws, A., Gu, Y., and Marshall, V. (2001). Biosynthesis, characterization, and design of bacterial exopolysaccharides from lactic acid bacteria. *Biotechnol. Adv.*, **19**, 597–625. DOI: 10.1016/ S0734-9750(01)00084-2.

98. Leathers, T. D. (2003). Biotechnological production and applications of pullulan. *Appl. Microbiol. Biotechnol.*, **62**, 468–473.

99. Leung, M. Y. K., Liu, C., Koon, J. C. M., and Fung, K. P. (2006). Polysaccharide biological response modifiers. *Immunol. Lett.*, **105**, 101–114.

100. Li, B., Lu, F., Wei, X., and Zhao, R. (2008). Fucoidan: Structure and bioactivity. *Molecules*, **13**, 1671–1695, DOI: 10.3390/molecules13081671.

101. Lim, S. N., Cheung, P. C. K., Ooi, V. E. C., and Ang, P. O. (2002). Evaluation of antioxidative activity of extracts from a brown seaweed, *Sargassum siliquastrum*. *J. Agri. Food Chem.*, **50**, 3862–3866. DOI: 10.1021/jf020096b.

102. Lima, J. A. L., Santos, N. D., Gomes, D. L., Cordeiro, S. L., Sabry, D. A., Costa, L. S., Freitas, M. L., Silva, N. B., Moura, C. E. B., Lemos, T. M. A. M., Leite, E. L., and Rocha, H. A. O. (2010). Evaluation of acute and subchronic toxicity of a non-anticoagulant, but antithrombotic algal heterofucan from the Spatoglossumschröederi in Wistar rats. *Braz. J. Pharmacognosy*, 1–6.

103. Liu, Z., Jiao, Y., Wang, Y., Zhou, C., and Zhang, Z. (2008). Polysaccharides-based nanoparticles as drug delivery systems. *Adv. Drug Deliv. Rev.*, **60**, 1650–1662.

104. Lothe, F. (1993). Polysaccharides and their derivatives. In: *Industrial Gums*, 3rd ed. (Whistler, R. L. and BeMiller, J. N., Eds.), Academic Press, San Diego.

105. MacLeod, A. M. and McCorquodale, H. (1958). Water-soluble carbohydrates of seeds of the *Gramineae*. *New Phytologist*, **57**(2), 168–182.

106. Mancilla-Margalli, N. A. and López, M. G. (2006). Water-soluble carbohydrates and fructan structure patterns from *Agave* and *Dasylirion* species. *J. Agri. Food Chem.*, **54**(20), 7832–7839.

107. Manzano, M., Citterio, B., Rondini, G. and Bertoldi, M. (1995). Microbiological aspects of natural starter in Montasio cheesemaking. *Ann. Microbiol.*, **1992**(42), 163–170.

108. Markager, S., Olesen, B., Arias, C., and Jensen, P. D. (2011). Bioenergy potential of *Ulva lactuca*: Biomass yield, methane production and combustion. *Biores. Technol.*, **102**, 2595–2604.

109. May, C. D. 2000. Industrial pectins: Sources, production and applications. *Carbohy. Polym.*, **12**(l), 79–99.

110. McClure, M. O., Moore, J. P., Blanc, D. F., Scotting, P., Cook, G. M. W., Keynes, R. J., Weber, J. N., Davies, D., and Weiss R. A. (1992). Investigations into the mechanism by which sulphated polysaccharides inhibit HIV infection in vitro. *AIDS Res. Hum. Retroviruses*, **8**, 19–26.

111. McHugh, D. J. (1987). Production, properties and uses of alginates. In: *Production and Utilization of Products from Commercial Seaweeds* (McHugh, D. J., Ed.), 288, 189, FAO Fish. Tech, Rome.

112. Miller-Chou, B. A. and Koenig, J. L. (2003). A review of polymer dissolution. *Prog. Polym. Sci.*, **28**(8), 1223–1270.

113. Mishra, A. and Malhotra, A. V. (2009). Tamarind xyloglucan: A polysaccharide with versatile application potential. *J. Mater. Chem.*, **19**, 8528–8536.

114. Mishra, M. U. and Khandare, J. N. (2007). Tamarind seed polysaccharides: Biodegradable polymer for colonic drug delivery. In: *Second International Conference and Indo-Canadian Satellite Symposium on Pharmaceutical Science, Technology, Practice and Natural Products,* Conference Chronicle, DDS, pp. 206.

115. Moller, I., Sørensen, I., Bernal, A. J., Blaukopf, C., Lee, K., Øbro, J., Pettolino, F., Roberts, A., Mikkelsen, J. D., Knox, J. P., Bacic, A., and Willats, W. G. (2007). High-throughput mapping of cell wall polymers within and between plants using novel microarrays. *Plant J.*, **50**, 1118–1128.

116. Morris, V. J. (1995). Bacterial polysaccharides. In: *Food Polysaccharides and Their Applications* (Stephen, A. M., Ed.), Marcel Dekker, New York, 341–375.

117. Morris, V. J. and Chilvers, G. R. (1984). Cold setting alginate-pectin mixed gels. *J. Sci. Food Agri.*, **35**(12), 1370–1376.

118. Mulloy, B., Mourão, P. A. S., and Gray, E. (2000). Structure/function studies of anticoagulant sulphated polysaccharides using NMR. *J. Biotechnol.*, **77**, 123– 135.

119. Nagaoka, M., Shibata, H., Kimura-Takagi, I., Hashimoto, S., Kimura, K., Makino, T., Aiyama, R., Ueyama, S., and Yokokura, T. (1999). Structural study of fucoidan from *Cladosiphon okamuranus. Glycoconj. J.*, **16**(1), 19–26.

120. Nakatsubo, F., Kamitakahara, H., and Hori, M. (1996). Cationic ring-opening polymerization of 3,6-di-O-benzyl-α- D-glucose 1,2,4 -orthopivalate and the first chemical synthesis of cellulose. *J. Am. Chem. Soc.*, **118**(7), 1677–1681.

121. Necas, J. and Bartosikova, L. (2013). Carrageenan: A review. *Veterinarni Medicina*, **58**(4), 187–205.

122. Nikolić, G. S. and Cakić, M. D. (2011). Analysis of bioactive polygosaccharide-metal complexes by modern FTIR spectroscopy:

Copper complexes. In: *Fourier Transforms: New Analytical Approaches and FTIR Strategies*, InTech, 15–44.

123. Northcote, D. H. (1972). Chemistry of the plant cell wall. *Annu. Rev. Plant Physiol.*, **23**, 113–132.

124. Novosel'skaya, I. L., Voropaeva, N. L., Semenova, L. N., and Rashidova, S. Sh. (2000). Trends in the science and applications of pectins. *Chem. Nat. Compd.*, **36**, 1–10.

125. O'Neill, M. A., Silvendran, R. R., and Morris, J. (1983). Structure of extracellular gelling polysaccharide produced by *Pseudomonas elodea. Carbohydr. Res.*, **124**, 123–133.

126. Oakenfull, D. G. (1991). The chemistry of high-methoxyl pectins. In: *The Chemistry and Technology of Pectin* (Walter, R. H., Ed.), Academic Press, New York.

127. Palaniraj, A. and Jayaraman, V. (2011). Production, recovery and applications of xanthan gum by *Xanthomonas campestris. J. Food Eng.*, **106**(1), 1–12.

128. Patankar, M. S., Oehninger, S., Barnett, T., William, R. J., and Clark, G. F. (1993). A revised for fucoidan may explain some of its biological activities. *J. Biol. Chem.*, **268**, 21770–21776.

129. Paterson, I. and Anderson, E. A. (2005). The renaissance of natural products as drug candidates, *Science*, **310** (5747), 451–453.

130. Pauly, M. and Keegstra, K. (2008). Cell-wall carbohydrates and their modification as a resource for biofuels. *Plant J.*, **54**, 559–568.

131. Pauly, M., Gille, S., Liu, L., Mansoori, N., De Souza, A., Schultink, A., and Xiong, G. (2013). Hemicellulose biosynthesis. *Planta*, **238**, 627–642, DOI: 10.1007/s00425-013-1921-1.

132. Payen, A. (1838). Mémoire sur la composition du tissu propre des plantes et du ligneux (Memoir on the composition of the tissue of plants and of woody [material]), *Comptes rendus*, **7**, 1052–1056.

133. Payen, A. (1842). Troisième mémoire sur le development végétaux'Extrait des mémoires de l'Academie Royale des Sciences: Tomes III des Savants Éntranges, Imprimerie Royale, Paris.

134. Pereira, M. S., Melo, F. R., and Mourao, P. A. S. (2002). Is there a correlation between structure and anticoagulant action of sulfated galactans and sulfated fucans? *Glycobiology*, **12**(10), 573–580.

135. Pereira, M. S., Mulloy, B., and Mourão, P. A. S. (1999). Structure and anticoagulant activity of sulfated fucans. *J. Biol.Chem.*, **274**, 7656–7667.

136. Prajapati, V. D., Jani, G. K., Zala, B. S., and Khutliwala, T. A. (2013). An insight into the emerging exopolysaccharide gellan gum as a novel polymer. *Carbohyd. Polym.*, **93**, 670–678.

137. Pszczola. D. E. (1993). Gellan gum wins 1FT's Food Technology Industrial Achievement Award. *Food Technol.*, **47**, 94–96.

138. Quinten, T. (2010). Evaluation of injection molding as a pharmaceutical production technology for sustained-release matrix tablets, D Pharm Thesis, 1–241.

139. Reis, R. A., Tischer, C. A., Gorrin, P. A. J., and Iacomini, M. (2002). A new pullulan and a branched (1-3)-, (1-6)-linked β-glucan from the lichenised ascomycete *Teloschistes flavicans*. *FEMS Microbiol. Lett.*, **210**, 1–5.

140. Rhein-Knudsen, N., Ale, M. T., Ajalloueian, F., and Meyer, A. S. (2017). Characterization of alginates from Ghanaian brown seaweeds: *Sargassum* spp. and *Padina* spp. *Food Hydrocolloid.*, **71**, 236–244, DOI: 10.1016/j.foodhyd.2017.05.016.

141. Richards, H. L., Baker, P. G. L., and Iwuoha, E. (2012). Metal nanoparticle modified polysulfone membranes for use in wastewater treatment: A critical review. *J. Surf. Eng. Mater. Adv. Technol.*, **2**, 183–193, http://dx.doi.org/10.4236/jsemat.2012.223029.

142. Ritsema, T. and Smeekens, S. (2003). Engineering fructan metabolism in plants. *J. Plant Physiol.*, **160**, 811–820.

143. Roberfroid, M. B. (2000). Prebiotics and probiotics: Are they functional foods? *Am. J. Clin. Nutr.*, **71**, 1682S–1687S.

144. Rolando, M. and Valente, C. (2007). Establishing the tolerability and performance of tamarind seed polysaccharide (TSP) in treating dry eye syndrome: Results of a clinical study. *BMC Ophthalmol.*, **7**, 5.

145. Rorrer, G. L. and Hawley, M. C. (1993). Vapor-phase HF solvolysis of cellulose: Modification of the reversion oligosaccharide distribution by in situ methanolysis, *Carbohydr. Polym.*, **22**(1), 9.

146. Rowell, R. M., Pettersen, R., Han, J. S., Rowell, J. S., and Tshabalala, M. A. (2005). Cell wall chemistry. In: *Handbook of Wood Chemistry and Wood Composites*, CRC Press, Florida, USA, 35–74.

147. Sabra, W. and Zeng, A. P. (2009). Microbial production of alginates: Physiology and process aspects. In: *Alginates: Biology and Applications, Microbiology Monographs* (Rehm, B. H. A., Ed.), vol. 13, 153–173, DOI: 10.1007/978-3-540-92679-5_7.

148. Sanderson, G. R. (1982). The interactions of xantham gum in food systems. *Prog. Food Nutr. Sci.*, **6**, 77–87.

149. Saravanakumar, G., Jo, D. G., and Park, J. H. (2012). Polysaccharide based nanoparticles: A versatile platform for drug delivery and biomedical imaging. *Curr. Med. Chem.*, **19**, 3212–3219.

150. Sarossy, Z., Blomfeldt, T. O. J., Hedenqvist, M. S., Kochk, C. B., Ray, S. S., and Plackett, D. (2012). Composite films of arabinoxylan and fibrous sepiolite: Morphological, mechanical, and Baffler properties. *ACS Appl. Mater. Inter.*, **4**, 3378–3386.

151. Satturwar, P. M., Fulzele, S. V., and Dorle, A. K. (2003). Biodegradation and in vivo biocompatibility of rosin: A natural film-forming polymer. *AAPS Pharm. Sci. Tech.*, **4**, 1–6.

152. Sawada, S. I., Sasaki, Y., Nomura, Y., and Akiyoshi, K. (2011). Cyclodextrin-responsive nanogel as an artificial chaperone for horseradish peroxidase. *Colloids Polym. Sci.*, **289**, 685–691.

153. Scheller, H. V. and Ulvskov, P. (2010). Hemicelluloses. *Annu. Rev. Plant Biol.*, **61**, 263–289.

154. Senthilkumar, K., Manivasagan, P., Venkatesan, J., and Kim, S. K. (2013). Brown sea-weed fucoidan: Biological activity and apoptosis, growth signaling mechanism in cancer. *Int. J. Biol. Macromol.*, **60**, 366–374.

155. Shahidi, F. and Rahman, M. J. (2018). Bioactives in seaweeds, algae, and fungi and their role in health promotion. *J. Food Bioact.*, **2**, 58–81.

156. Shingel, K. I. (2004). Current knowledge on biosynthesis, biological activity, and chemical modification of the exopolysaccharide, pullulan. *Carbohydr. Res.*, **339**, 447–460.

157. Sigma-Tau Industrie Farmaceutiche Riunite, S. P. A., British Patent 1,206,870, 1970; *Chem. Abstr.,* 73,133985d, 1970.

158. Silva, T. H, Alves, A., Ferreira, B. M., Oliveira, J. M., Reys, L. L., Ferreira, R. J. F., Sousa, R. A., Silva, S. S., Mano, J. F., and Reis, R. L. (2012). Materials of marine origin: A review on polymers and ceramics of biomedical interest. *Int. Mater. Rev.*, **57**, 276–306.

159. Singh, R. S. and Kaur, N. (2015). Microbial biopolymers for edible film and coating applications. In: *Advances in Biotechnology* (Nawani, N. N., Khetmalas, M., Razdan, P. N., and Pandey, A., Eds.), IK International Publishing, India, 187–216.

160. Singh, R. S., Kaur, N., and Kennedy, J. F. (2015). Pullulan and pullulan derivatives as promising biomolecules for drug and gene targeting. *Carbohydr. Polym.*, **123**, 190–207.

161. Smouter, H. and Simpson, R. J. (1991). Fructan metabolism in leaves of *Lolium rigidum* Gaudin. *New Phytologist*, **119**(4), 509–516.

162. Sørensen, H. R., Meyer, A. S., and Pedersen, S. (2003). Enzymatic hydrolysis of water-soluble wheat arabinoxylan. 1. Synergy between α-L-arabinofuranosidases, endo-1,4-β-xylanases, and β-xylosidase activities. *Biotechnol. Bioeng.*, **81**(6), 726–731.

163. Sriamornsak, P. (2001-2002). Pectin: The role in health. *J. Silpakorn University*, **21–22**, 60–77.

164. Stanford, E. C. C. (1883). On algin: A new substance obtained from some of the commoner species of marine algae. *Chem. News*, **47**, 254–257.

165. Stanford, E. C. C. (1884). On the economic applications of seaweeds. *J. Soc. Arts*, **32**, 717–732.

166. Staudinger, H. (1960). Die hochmolekularen organichen Verbindungen: Kautschuk und Cellulose, 2nd edition, Springer Verlag, Berlin.

167. Sundarraj, A. A. and Ranganathan, T. V. (2017). A review: Pectin from agro and industrial waste. *Int. J. Appl. Env. Sci.*, **12**(10), 1777–1801.

168. Sundarraj, A. A., Rubila, S., Jayabalan, R., and Ranganathan, T. V. (2012). A review on pectin: Chemistry due to general properties of pectin and its pharmaceutical uses. *Open Access Sci. Rep.*, **1**(12), 1–4.

169. Sutherland, I. W. (1996). Extracellular polysaccharides. In: *Biotechnology* (Rehm, H. J. and Reed, G., Eds.), 2nd edition, Vol. 6, VCH, Weinheim, 613–657.

170. Technical Evaluation Report. (2015). Compiled by OMRI for the USDA National Organic Program, 1-23.

171. Thakur, B. R., Singh, R. K., and Handa, A. K. (1997). Chemistry and uses of pectin: A review. *Crit. Rev. Food Sci. Nutr.*, **37**, 47–73.

172. Torres, M. L., Cortizo, A. M., Oberti, T. G., and Fernández, J. M. (2016). Characterization of commercial and algae (*Undaria pinnatifida*) extracted sodium alginate for future application in bone tissue engineering, 1–13.

173. Tungland, B. (2018). Nondigestible fructans as prebiotics. In: *Human Microbiota in Health and Disease: From Pathogenesis to Therapy*, Academic Press, 349–379, https://doi.org/10.1016/b978-0-12-814649-1.00008-9.

174. Umapoorani, T., Periyanayagam, K., Indumathi, S., Balasubramanian, R., Sugithra, B., Subbulekshmi, N., and Velmurugan, R. (2016). Antimitotic activities of ulvan from *Ulva lactuca* (L) by using *Allium cepa* root tip model. *Indo Am. J. Pharm. Res.*, **6**(6), 5779–5783.

175. Valent, B. S. and Albersheim, P. (1974). The structure of plant cell walls: V. On the binding of xyloglucan to cellulose fibers. *Plant Physiol.*, **54**, 105–108.

176. van de Velde, F., Lourenco, N. D., Pinheiro, H. M., and Bakkerd, M. (2002). Carrageenan: A food-grade and biocompatible support for immobilisation techniques. *Adv. Syn. Catal.*, **344**, 815–835.

177. van den Ende, W., de Coninck, B., and van Laere, A. (2004). Plant fructan exohydrolases: A role in signaling and defense? *Trends Plant Sci.*, **9**(11), 523–528.

178. van den Ende, W., Michiels, A., de Roover, J., and van Laere, A. (2002). Fructan biosynthetic and breakdown enzymes in dicots evolved from different invertases. Expression of fructan genes throughout chicory development. *Sci. World J.*, **2**, 1273–1287. doi:10.1100/tsw.2002.288.

179. van den Ende, W., Mintiens, A., Speleers, H. Onuoha, A. A., and van Laere, A. (1996). The metabolism of fructans in roots of *Cichorium intybus* during growth, storage and forcing. *New Phytologist*, **132**(4), 555–563.

180. van der Meulen, R., Avonts, L., and de Vuyst, L. (2004). Short fractions of oligofructose are preferentially metabolized by *Bifidobacterium animalis* DN-173 010. *Appl. Environ. Microbiol.*, **70**, 1923–1930.

181. Vanzin, G. F., Madson, M., Carpita, N. C., Raikhel, N. V., Keegstra, K., and Reiter, W.-D. (2002). The *mur2* mutant of *Arabidopsis thaliana* lacks fucosylated xyloglucan because of a lesion in fucosyltransferase AtFUT1. *Proc. Natl. Acad. Sci.*, **99**, 3340–3345.

182. Verbruggen, M. A., Spronk, B. A., Schols, H. A., Beldman, G., Voragen, A. G. J., Thomas, J. R., and Kamerling, J. P. (1998). Structures of enzymically derived oligosaccharides from sorghum glucuronoarabinoxylan. *Carbohydr. Res.*, **306**, 265–274.

183. Vergara, C. E. and Carpita, N. C. (2001). β-D-glycan synthases and the *CesA* gene family: Lessons to be learned from the mixed-linkage (1→3,1→4)β-D-glucan synthase. *Plant Mol. Biol.*, **47**, 145–160.

184. Vergauwen, R., van den Ende, W., and van Laere, A. (2000). The role of fructan in flowering of *Campanula rapunculoides*. *J. Exp. Botany*, **51**(348), 1261–1266.

185. Vidal, B. C., Haroun, F., Ellouali, M., Blondin, C., Fischer, A. M., Agostini, A., and Jozefonvicz, J. (1995). Biological activities of polysaccharides from marine algae. *Drugs Future*, **20**, 1237–1249.

186. Vidal, B. C., Chaubet, F., Chevolot, L., Sinquin, C., Theveniaux, J., Millet, J., Sternberg, C., Mulloy, B., and Fischer, A. M. (2000). Relationship

between antithrombotic activities of fucans and their structure. *Drug Dev. Res.*, **51**, 216–224.

187. Vilhalva, D. A. A., Cortelazzo, A. L., Carvalho, A. L. M. A. M., and Figueiredo-Ribeiro, R. D. C. L. (2011). Histochemistry and ultrastructure of *Campuloclinium chlorolepis* (Asteraceae) tuberous roots accumulating fructan: Evidences of functions other than reserve carbohydrate. *Aus. J. Botany*, **59**(1), 46–52.

188. Viswanathan, S. and Nallamuthu, T. (2014). Extraction of sodium alginate from selected seaweeds and their physiochemical and biochemical properties. *Int. J. Innov. Res. Sci. Eng. Technol.*, **3**(4), 10998–11003.

189. Weiner, M. L. (2014). Food additive carrageenan: Part II: A critical review of carrageenan in vivo safety studies. *Crit. Rev. Toxicol.*, **44**(3), 244–269, DOI: 10.3109/10408444.2013.861798.

190. Weiner, M. L. and Lois, A. K. (2000). *Excipient Toxicity and Safety*, Dekker, New York, p. 210.

191. Xu, Q., Yajimaw, T., Liw, W., Saitow, K., Ohshima, Y., and Yoshikaiw, Y. (2006). Levan (β-2,6-fructan), a major fraction of fermented soybean mucilage, displays immunostimulating properties via toll-like receptor 4 signalling: Induction of interleukin-12 production and suppression of T-helper type 2 response and immunoglobulin E production. *Clin. Exp. Allergy*, **36**, 94–101.

192. Young, R. (1986). *Cellulose Structure Modification and Hydrolysis*, Wiley, New York, ISBN 0471827614.

193. Yuen, S. (1974). Pullulan and its applications. *Process Biochem.*, **9**, 7–9.

194. Zhang, Q., Li, Z., Xu, Z., Niu, X., and Zhang, H. (2003). Effects of fucoidan on chronic renal failure in rats. *Planta. Med.*, **69**, 537–541.

Chapter 2

Pullulan: Recent Progress and Technological Prospects

Tanvir Arfin
Environmental Materials Division, CSIR-NEERI, Nehru Marg, Nagpur, Maharashtra, India
tanvirarfin@gmail.com

According to the literature survey of recent years, pullulan has gathered interest from various researchers. It is a polysaccharide possessing excellent properties because of the unique glycosidic linkage. It is modified chemically for reducing the solubility of water, enhancing the sensitivity of pH, and introducing reactive functional groups, which enable it to be used in various applications. It is employed for surface modification of polymeric material to make it highly compatible with blood as well as bioinert. Since pullulan is adhesive, it is applied for the application of wound dressing. It can be used in various ways due to underlying features such as thermal stability and elastic property. It forms the compression molding, and it also gets decomposed at high temperature without any formation of toxic gases. Such a feature can be applied with biological characteristics for developing implant devices. In this chapter, it

Pullulan: Processing, Properties, and Applications
Edited by Shakeel Ahmed and Aisverya Soundararajan
Copyright © 2021 Jenny Stanford Publishing Pte. Ltd.
ISBN 978-981-4877-17-6 (Hardcover), 978-1-003-05694-2 (eBook)
www.jennystanford.com

is clarified that pullulan, a unique polymer, is used exclusively for biomedical application since its characteristics are tapped to possess high potential.

2.1 Introduction

At present, in the research study, the development of society is an essential topic for sustainable development. The concept of the studies has mainly gathered interest due to advancement and prospect of regeneration since it helps in maintaining equilibrium in using the resources that are linked together with human beings and the ecosystem [1–5]. In the recent era of scientific research, it has been observed that the consolidation of innovative materials encapsulated with advanced development is required for excellent and favorable outcome [6–10]. The developed material is used in different classes of analytical chemistry for the quantitative and qualitative discerning of the least sample present in a complicated way [11–15]. The constituents are adequate to carry out research at full pace and more widely and efficiently in the techno-economic process [16–20]. We are aware that polymers are highly stable, so the degradation cycles are indefinite within the biosphere. There have been issues related to pollution, mainly environmental pollution, which is caused by synthetic polymers, namely, waste plastic and water-soluble polymers. Polymers and plastics are significant components for the better survival of human beings every day. Yet because of the stability and resistance toward degradation, all these components accumulate within the environment at 8% by weight and 20% by volume of the disposal area [21–25]. Polymers are a group of giant molecules comprising distinct building blocks that are connected, forming long chains. The simple blocks are known as monomers, and complicated ones are called repeat units. Biopolymers are polymers that are formed in natural conditions during the growth cycle of all living organisms and are, therefore, also called natural polymers. They form within living cells from significant and complex hydrocarbons. Moreover, special attention is given to complex hydrocarbon polymers developed from bacteria and fungi, especially polysaccharides such as chitin, xanthan, pullulan, and so on [26–30]. The focus of the research study is to produce novel biodegradable polymers. Hence to fulfil the demands, many biodegradable polymers are synthesized at the phase of growth cycles of all living organisms.

Biopolymers obtained from various microorganisms possess unique and extraordinary features. Owing to such features, they are applied in multiple fields such as the food and pharmaceutical sectors. The application of biopolymers involves a binding agent. The main uses are in the form of stabilizers, emulsifiers, binders, and lubricants. The technological advancement is because of the exploitation of properties generated through polymeric substances such as composite, blends, and so on. Blends of polymers are synthesized by the combination of two or more polymers. As a result, a blending system is obtained, which is a superior property in comparison to single polymers. Pullulan is a polymeric yield of the fermentation medium of black yeast *Aureobasidium pullulans*. Pullulan was first proposed in 1959 for nominating extracellular polysaccharide, which is water soluble, prepared from *A. pullulans* [31]. In 1938, Bauer was the first to observe polysaccharides [32]. Commonly known as *Pullularia pullulans*, it is a fungus found in untreated water or sewage [33].

Aureobasidium (a genus) mainly consists of a membrane capable of surviving in varied types of habitats. *A. pullulans* is found in phyllosphere habitats such as grapevines, including high morphology and genetic diversity [34]. *A. pullulans*, a species of yeast, gets isolated from berries of grapes in the maturity stage [35] and also from tissues of vines developed from healthy as well as diseased vines [36].

The utilization of pullulan is gathering interest in the form of suitable polymeric material possessing competitive and economic features along with natural gums generated from marine algae [37].

2.2 Historical Outline

Pullulan is used in Japan in different food ingredients as well as a pharmaceutical bulking source. It is used primarily as glazing agents, which display oxygen barrier properties [38]. According to the application, pullulan is considered a suitable agent in GRAS status in the United States. Some recent studies have reported abdominal fullness with a dose of 10 gm dose/day, and in accordance with the gastrointestinal symptoms, this was considered as a high dose. In pharmaceutical tables, it is mentioned as an excipient, and in Japanese standards, it is known as ingredient of drugs. [39]. The production of pullulan was the outcome of starch syrup during 1883, but the film was commercialized in 1982 by Hayashibara.

2.3 Pullulan Biosynthesis

2.3.1 Mechanism of Pullulan Biosynthesis

Pullulans are intracellular polymers synthesized within the membranes of cell wall, which are then secreted to the outer surface of the cell in the form of a loose, slimy layer [40]. As such various investigations have been carried out on cytological and physiological characteristics of *A. pullulans.* The mechanism related to biosynthesis is not totally understood till date. Duan et al. [41] have suggested the pathway for the synthesis of pullulan. The glucose units obtained as a result require the availability of three essential enzymes that convert into pullulan. They include α-phosphoglucose mutase, UDPG-pyrophosphorylase, and glycosyltransferase.

A. pullulans is converted into various sources of carbon into precursor UDPG by using hexokinase and isomerase. Pullulan can also be synthesized by sucrose along with the cell-free enzymes of *A. pullulans* only when ATP and UDPG are available in the mixture of the reaction [42]. ADPG cannot act as a substitute for UDPG, which clarifies that pullulan chains and precursors originate from UDPG.

Figure 2.1 shows the pathway for pullulan synthesis. Even so, the precise mechanism of synthesis of pullulan has not been appropriately comprehended till data because of cytological features and complex physiology of the microorganism [43].

Hence, the proposed pathway for biosynthesis of pullulan can be suitable for metabolic activity.

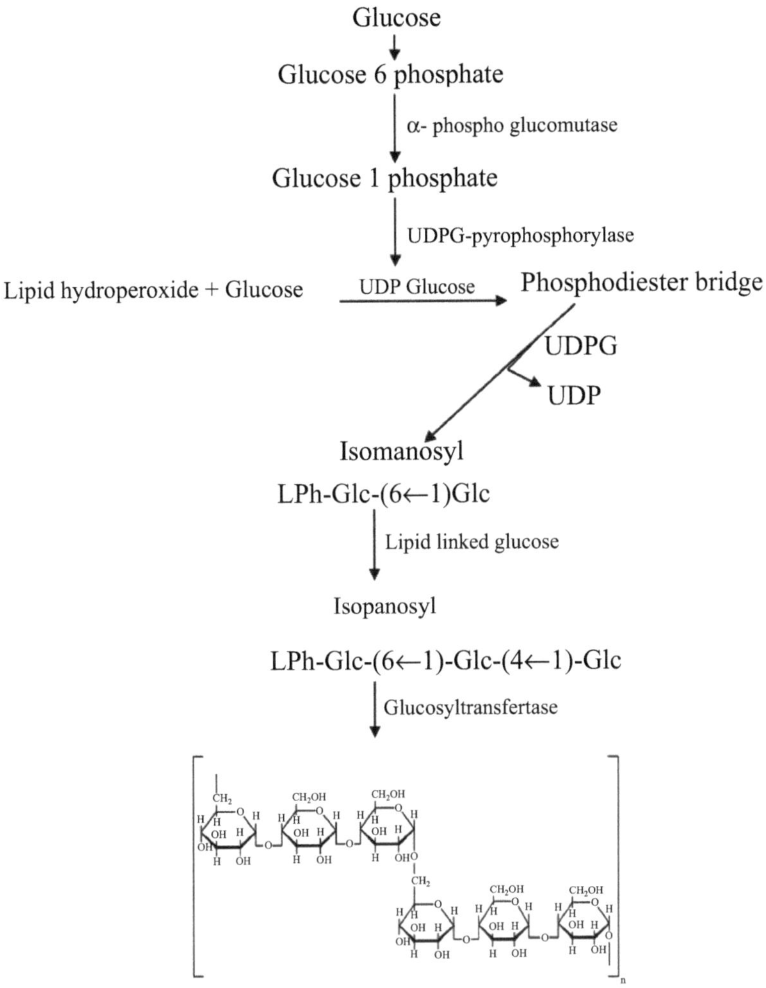

Figure 2.1 A schematic representation of the biosynthesis path of pullulan fabrication (reprinted with permission from Ref. [44] © Elsevier).

2.4 Products

Cooke suggested in his work the presence of *A. pullulans* in 1891 [45]. As such, it could be relevant enough and helpful for researchers to collect information about the physiology and life of fungus. The broad products are isolated, characterized, and also tested for different biological as well as non-biological functions. The antimicrobials, siderophores, and enzymes are the prominent products.

2.4.1 Antimicrobials

Bacteria and yeast exhibit mutual behavior for each other for colonizing the surface of berries. Due to interaction, they tend to enhance the nutrients-capture capabilities, resist environmental stress, and interact with another category of microorganisms, namely molds and viruses. There are other species that show antagonistic effect for the development of mold. *A. pullulans* exhibits the underlying antagonistic properties for yeast as well as fungi. They are speculated to completely enhance grape ecology [35].

2.4.2 Enzymes

As such, enzymes are capable of performing the task of antimicrobials, but enzyme production and activity need to play significant jobs for different biotechnological applications. *A. pullulans* produces various enzymes such as amylases [46], cellulose [47], lipases [48], xylanases [49], proteases [50], laccase [51], and mannanase [52]. At present enzymes, namely, pectinases, glucanases, xylanases, and proteases, are employed in the wine industry for enhancing the technique of clarification and processing. Baffi et al. [53] have reported the capabilities of non-Saccharomyces yeasts and their role in the aroma of wine as they got to have a notable increase in monoterpenes.

2.4.3 Pullulan and Other Polysaccharides

A. pullulans products are unbranched and extracellular polysaccharides pullulan comprising α-(1→6) linkages of α-(1→4)-linked maltotriose units as shown in Fig. 2.2 [54]. The obtained elastic and sticky polymer helps in the formation of oxygen-impermeable film, possessing property required for elaborating the idea of the availability of various anaerobic bacteria on the surface of berries. Hence, for the preservation of the cell, the microorganisms need nutritive sources. In this context, biofilms can be considered for nutrient trapping [55].

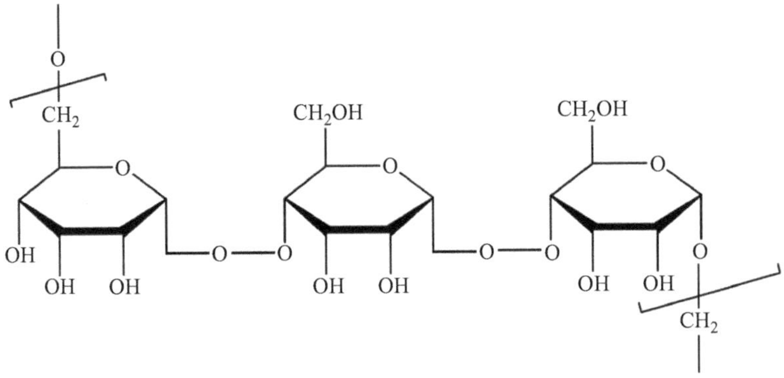

Figure 2.2 Structure of pullulan.

The production of pullulan by the fungus during the film's initial fermentation steps helps stabilize the process, and it improves the feeling of the wine because of the rheological properties of molecules in aqueous and ethanolic media. The polysaccharides tend to grow the aroma, which, on the other hand, enhances the delivery as well as the perception because of the physicochemical reaction that occurs along with the aromatic compounds. It has also been observed that it is good enough to retain the dominate color and antioxidant capability of red wine. Various types of microbial fermentation are as shown in Fig. 2.3.

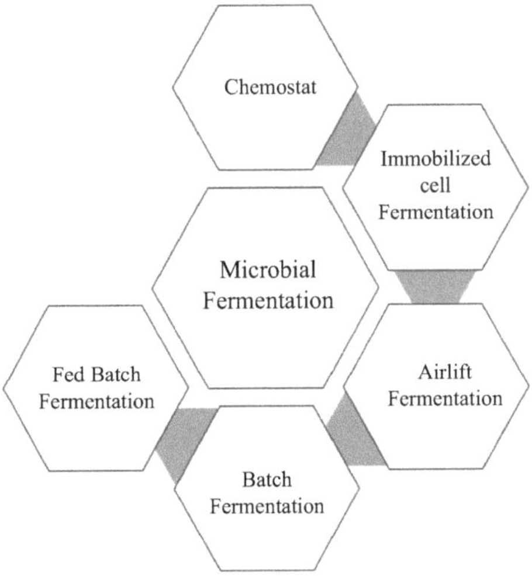

Figure 2.3 Different types of microbial fermentation.

In the past few years, different other characteristics of pullulan were also developed related to utility in the medical field, especially in the drug-delivery system and the connection of molecules to different cells, namely, cancer and liver cells [56]. The underlying features can be responsible for influencing the antioxidant property of red wine and helping in the investigation of the advantage of wine in human health. Li et al. [57] presented a review of *Aureobasiduim* sp. and its biosynthesis, regulation mechanism of the extracellular polymer. The most significant substrate to produce pullulan is grape skin pulp, commonly the hot water extract of skin pulp. The product obtained has high molecular weight and is relatively pure as well.

2.4.4 PMA

PMA is obtained as a natural biopolymer from various microorganisms, which mainly involve *A. pullulans*. It has different properties, such as water solubility, biodegradability, biocompatibility, and application in the pharmaceutical sector [58]. No report on its application has been suggested. However, useful relation can be developed through the wine waste in the form of substrate for the production of PMA and also provide coating on grapes for post-harvest protection.

2.4.5 Liamocins

In 1994, Kurosawa et al. [59] investigated the manufacturing of heavy oil within the *Aureobasiduim sp.* culture medium. Liamocin could be responsible for exhibiting the biological activities in the form of antimicrobials and anticancer sources [60]. Liamocin plays a significant role in controlling diseases in plants, especially in grapes.

2.4.6 Siderophores

Siderophores are low molecular weight molecules with high affinity and chelating ability for iron. They are produced by microorganisms such as fungi and bacteria that reside in iron-depleted surroundings. They solubilize the Fe^{3+} and Fe^{2+} forms of iron and sequester them as well. Siderophore is a significant molecule for various applications such as agricultural, pharmaceutical, environmental, and biotechnological fields. In the literature survey, only one strain of *A. pullulans* is mentioned for the production of siderophores. The theoretical studies of *A. pullulans* suggest the effects of siderophores on the biological process to control pathogen of human beings and plants [61].

2.4.7 Agro-industrial Wastes

Pullulan is produced from the different substrates of carbohydrate integrated into defined or non-defined media. Various other agro-industrial wastes are favorable to produce pullulan. The use of substrates is both environmentally and economically beneficial because they are either cheap or have negative costing. The development of some region will be helpful for the creativity of small or medium enterprises (SME), which leads to job opportunity and safeguards the environment from the problems of different types of pollution. A separate application of agro-industries is listed in Fig. 2.4.

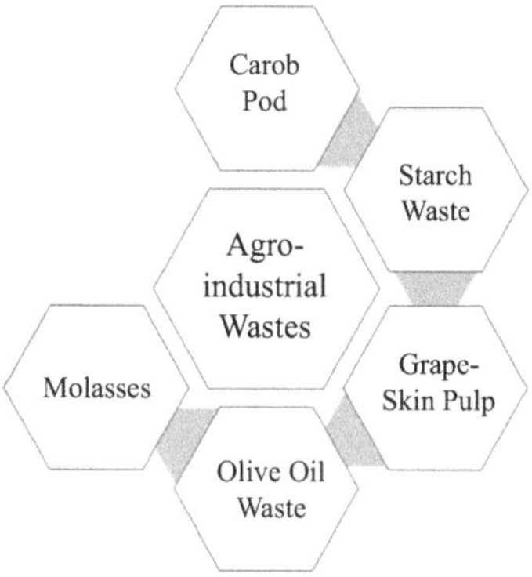

Figure 2.4 Various applications of agro-industries.

2.5 Disadvantages of Pullulan

Even if there are numerous applications of pullulan, it faces certain drawbacks that limit the utilization of pullulan in various industrial sectors. Pullulan is too expensive and brittle [62]. The hydrophilic nature tends to minimize the mechanical strength of pullulan. The main demerits are its inability to provide a surface for supporting cell adhesion and spreading during the process of cell proliferation and osteogenesis for bone tissue engineering [63]. Pullulan does not possess any of the antimicrobial properties as reported to date [64].

2.6 Aromatic Features

According to environmental importance, microorganisms are mainly grouped into three classes: (i) species that can be controlled easily without spoiling the wine during the manufacturing process, (ii) species that are fermented to convert sugar and malic acid, and (iii) species that spoil [65]. *A. pullulans* is holding a superior place in studies of grapevine terroirs, enabling it to be kept in the first group. It gives flavor to red wine [66]. There is no report of endophytic microorganisms playing any role in aromatic grape compounds. Studies based on grapevine endophyte have been progressing, and this will develop in the future generation [67].

2.7 Factors

Different factors influence the production of biopolymer. The ratio of carbon to nitrogen (C/N) of 10 is suitable for producing exopolysaccharide in the submerged process of fermentation. During the solid-state fermentation technique, the moisture content and size of particle influence the creation of pullulan. A different application of agro-industries is listed in Fig. 2.5.

2.8 Semi-synthetic Derivatives

To influence its activities and broaden its applications in various fields, pullulan can be derived very conveniently from the following different methods.

2.8.1 Chemical Modification

Pullulan's chemical structure can be varied by using the grafting technique. Its structure is composed of repeating units of nine hydroxyl groups, which can be used for substitution reactions. Its use can be boosted by using the grafting method to substitute the hydroxyl groups with various chemical groups. The relativity of the hydroxyl group is mainly based on the polarity of solvent and reagents. The chemical modification approaches are shown in Fig. 2.6.

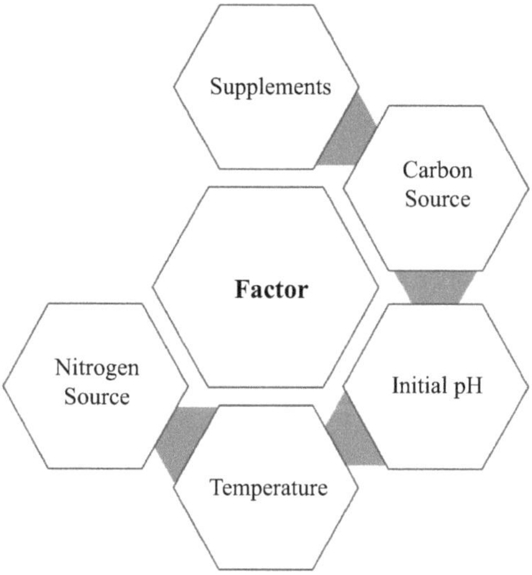

Figure 2.5 Influence of factors.

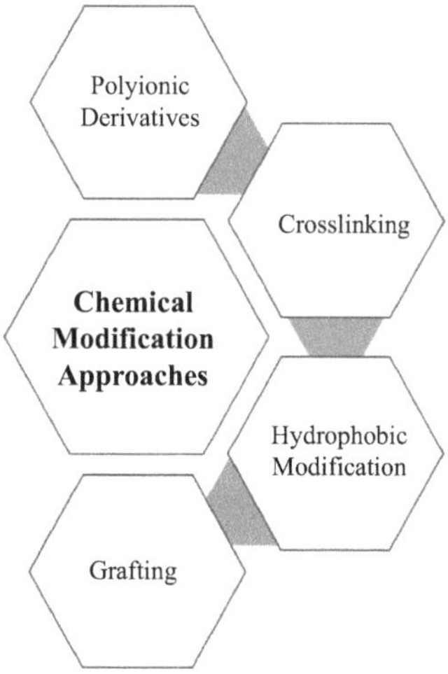

Figure 2.6 Various chemical modification approaches.

The chemical modification process involves different steps, such as:

2.8.1.1 Carboxymethyl pullulan

Carboxymethylation is the reaction performed on neutral polysaccharide for permitting the further process of chemical modification and also to assist in the mechanism of solubility in aqueous solution.

2.8.1.2 Sulfation

Pullulan is derived through the reaction along with sulfur to promote the best suitable alternative for heparin.

2.9 Applications

Various industrial and commercial uses of pullulan have been suggested to date [68]. In this context, it is preferred that there are several patents indicating its vast applications.

2.9.1 Food

Pullulan, a polysaccharide polymer, is tasteless and odorless. It is impermeable to oxygen and is soluble in water; its membrane is employed for coating and is used as packaging materials for various food products and drugs [69]. It can also be applied in the form of partial substitute for starch in low-calorie food as it does not degrade through *in vivo* digestive enzymes such as amylase and pepsin. It is used as microencapsulating material for spices and flavors to be used as seasonings [70]. The oxygen-resistance ability of pullulan films helps protect the freely oxidized fats and vitamins in food substances. The films are used as coating or packaging substance in the foods that are dry, such as nuts, noodles, meats, vegetables, and confectionary items [71]. Pullulan does not get comprehended from bacteria and molds. So it is mixed with different food or non-food substances for influencing the shelf-life of any products to protect the contents from degradation, as shown in Table 2.1. Pullulan is added in the process of making cakes as it helps to retain shape, moisture, and even the look of the cake after preparation. In frozen foods, it helps in preventing the loss of water or drip, and it also tends to improve the yield of production in other food products as well.

Table 2.1 Applications in food industries

Role	Uses	Reference
Dried food packaging	Confectionaries	[72]
Stabilizer	Emulsion	[73]
Preservative	Frozen foods	[74]
Starch substitution	Baked goods	[75]
Binder	Plant fertilizer	[76]

2.9.2 Biomedicine

Pullulan has various biomedical applications because of its extraordinary structural features. Figure 2.7 shows the biomedical use of pullulan. Modified pullulan derivatives are analyzed only for *in vitro* studies. In research studies, the focus is primarily on the *in vivo* ability of pullulan and its derivatives, and cationized pullulan is one such derivative that has been established as an efficient gene carrier for *in vivo* studies. The bulk and surface properties of the biomaterial applied in medical implants are influenced directly and allow for controlling the aggressive interaction occurring at the interface of tissue implant. The alteration in the properties of biomaterials occurs *in vivo* and is known to influence their designing for significant applications. Pullulan is employed to target the tumor cell and is also used for gene therapy. The endocytic pathway mediates the gene-delivery process. Gene therapy is executed by making the use of a virus, but the main demerit is that the virus is immunogenic, causes diseases, and is even dangerous. Diethylenetriamine penta acetic acid (DTPA) is employed in magnetic resonance imaging of hepatocytes in the treatment of cancer and is also recognized as a good substitute for dextrans in blood plasma [77]. A polymer with high water solubility can be employed as a plasma expander, so pullulan is favorable as it is a water-soluble polymer. It has also been revealed that pullulan is used in the form of plasma expander up to the molecular weight of nearly 60 kDa. Pullulan is cast with the beads that are microporous and applied for mineralization in bone deformities and repair. These beads can be further used in generating new fill-up material to repair bone defects of different sizes. All such developments suggest that novel innovation in the properties of pullulan may pave the way for its suitable application in biomedical studies and researches.

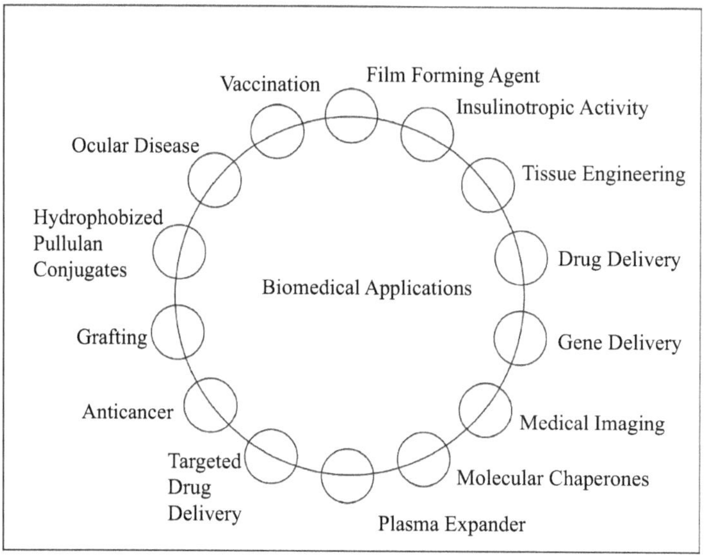

Figure 2.7 Various applications of pullulan in biomedical areas.

2.9.3 Water Remediation

The contamination of water from various sources, especially from industrial waste, aromatic molecules, dyes, and heavy metals, is growing tremendously and has become a threat to the world. Above all, the most dangerous one is the non-biodegradable heavy metals that have demerits such as toxic nature, carcinogenic, and mutagenic even if present in a small amount. Therefore, an efficient, highly effective treatment is required to treat such heavy metals that release effluents [78]. It has been observed that extracellular polysaccharides that yield microbial strains can be promising for water remediation because of the biosorption of heavy metals from the polysaccharide. Such a process works out when metal cations and acidic functional group of polymers interact with each other

[79]. During the process, the network of swollen macromolecules permits for the diffusion mechanism and helps in the physical entrapment of pesticides as well as pollutants. The high specificity and even the flocculation capacity are attained by employing the tailored polysaccharides [80]. The cationized pullulan showed high and efficient flocculation for pesticides when the pH of the emulsion was reduced, and the degree of substitution was raised [81].

2.10 Conclusion

Today, since the world is exploring the field of science, the material has gathered interest in different applications [82–86]. The materials possess varied functions in the commercial sector, which exhibit extraordinary results and impressive features [87–91]. Pullulan has generated interest in the field of research for many years because of its exceptional properties. It is an edible bio-polysaccharide possessing various applications in the pharmaceutical sector. The glycosidic linkage in pullulan is a significant property. It is manufactured through the process of fermentation, especially fermentation of coconut by-products, molasses of beet, and wastes of agro-industry. Pullulan is derived from chemical reaction. It is mostly employed in the biomedical and pharmaceutical sector for tissue engineering and grafting. It is applied in the target delivery of drugs to liver and cancer cells. Pullulan holds a high position in the food and pharmaceutical sectors. Pullulan possess various features as it can form compressive moldings, get decomposed at a high temperature of 250–280°C without the formation of any toxic gases. Such a feature of pullulan on interaction to the biological properties is used for the development of implant devices. Different factors are required for production practices, sales economy, and selecting proper substrate and strain of *A. pullulans*. On the other hand, it safeguards the environment from pollution.

Acknowledgments

The author acknowledges the Knowledge Resource Centre, CSIR-NEERI (CSIR-NEERI/KRC/2019/AUGUST/EMD/1) for their support.

References

1. Bushra, R., Arfin, T., Oves, M., Raza, W., Mohammad, F., Khan, M. A., Ahmad, A., Azam, A., and Muneer, M. (2016). Development of PANI/MWCNTs decorated with cobalt oxide nanoparticles towards multiple electrochemical, photocatalytic and biomedical application sites. *New J. Chem.*, **40**, pp. 9448–9459.
2. Arfin, T., Bushra, R., and Mohammad, F. (2016). Electrochemical sensor for the sensitive detection of o-nitrophenol using graphene oxide-poly(ethyleneimine) dendrimer-modified glassy carbon electrode. *Graphene Technol.*, **1**, pp. 1–15
3. Arfin, T. and Rafiuddin. (2009). Transport studies of nickel arsenate membrane. *J. Electroanal. Chem.*, **636**, pp. 113–122.
4. Arfin, T. and Rafiuddin. (2009). Electrochemical properties of titanium arsenate membrane. *Electrochim. Acta*, **54**, pp. 6928–6934.
5. Arfin, T. and Mohammad, F. (2014). Electrochemical, dielectric behavior and in vitro antimicrobial activity of polystyrene-calcium phosphate. *Adv. Ind. Eng. Manag.*, **3**, pp. 25–38.
6. Arfin, T. and Fatma, S. (2014). Synthesis, influence of electrolyte solutions on impedance properties and in-vitro antibacterial studies of organic-inorganic composite membrane. *Adv. Ind. Eng. Manag.*, **3**, pp. 19–30.
7. Arfin, T. and Fatima, S. (2014). Conductometric studies with polystyrene calcium phosphate membrane. *Asian J. Adv. Basic Sci.*, **2**, pp. 1–14.
8. Arfin, T. and Mohammad, F. (2013). DC electrical conductivity of nano-composite polystyrene-titanium-arsenate membrane. *J. Ind. Eng. Chem.*, **19**, pp. 2046–2051.
9. Mohammad, F. and Arfin, T. (2013). Cytotoxic effects of polystyrene-titanium-arsenate composite in cultured H9c2 cardiomyoblasts. *Bull. Environ. Contam. Toxicol.*, **91**, pp. 689–696.
10. Arfin, T. and Yadav, N. (2012). Impedance characteristics and electrical double-layer capacitance of composite polystyrene-cobalt-arsenate membrane. *J. Ind. Eng. Chem.*, **19**, pp. 256–262.
11. Arfin, T. and Rafiuddin. (2012). Metal ion transport through a polystyrene-based cobalt arsenate membrane: Application of irreversible thermodynamics and theory of absolute reaction rates. *Desalination*, **284**, pp. 100–105.

12. Arfin, T. and Yadav, N. (2012). Impedance characteristics and electrical double layer capacitance of polystyrene-based nickel arsenate membrane. *Anal. Bioanal. Electrochem.*, **4**, pp. 135–152.

13. Arfin, T. and Rafiuddin. (2011). An electrochemical and theoretical comparison of ionic transport through a polystyrene-based cobalt arsenate membrane. *Electrochim. Acta*, **56**, pp. 7476–7483.

14. Arfin, T., Jabeen, F., and Kriek, R. J. (2011). An electrochemical and theoretical comparison of ionic transport through a polystyrene based titanium-vanadium (1:2) phosphate membrane. *Desalination*, **274**, pp. 206–211.

15. Arfin, T. and Rafiuddin. (2010). Thermodynamics of ion conductivity of alkali halide across a polystyrene-based titanium arsenate, membrane. *Electrochim. Acta*, **55**, pp. 8628–8631.

16. Mohammad, F., Arfin, T., and Al-Lohedan, H. A. (2019). Biocompatible polylactic acid-reinforced nickel-arsenate composite: Studies of electrochemical conductivity, mechanical stability, and cell viability. *Mater. Sci. Eng. C*, **102**, pp. 142–149.

17. Arfin, T., Falch, A., and Kriek, R. J. (2012). Evaluation of charge density and the theory for calculating membrane potential for a nano-composite nylon-6,6 nickel phosphate membrane. *Phys. Chem. Chem. Phys.*, **14**, pp. 16760–16769.

18. Arfin, T., Bushra R., and Kriek, R. J. (2013). Ionic conductivity of alkali halides across a polyaniline-zirconium (IV)-arsenate membrane. *Anal. Bioanal. Electrochem.*, **5**, pp. 206–221.

19. Arfin, T. and Rangari, S. N. (2018). Graphene-oxide-ZnO nanocomposite modified electrode for the detection of phenol, *Anal Methods*, **10**, pp. 347–358.

20. Mohammad, F., Arfin, T., and Al-lohedan, H. A. (2019). Enhanced biosorption and electrochemical performance of sugarcane bagasse derived a polylactic acid-graphene oxide-CeO_2 composite. *Mater. Chem. Phys.*, 229, pp. 117–123.

21. Arfin, T. and Sonawane, K. (2018). *Bio-based Materials for Food Packaging: Green and Sustainable Advanced Packaging*, ed. Ahmed, S., Chapter 1 "Bio-based materials: past to future," (Springer Nature, Singapore), pp. 1–32.

22. Arfin, T. (2019). *Marine Polysaccharides: Advances and Multifaceted Application*, eds. Ahmed, S. and Soundarajan, A., Chapter 1 "Marine polysaccharides: An overview," (Jenny Stanford Publishing, Singapore), pp. 1–11.

23. Arfin, T. and Tarannum, A. (2017). *Green Polymeric Materials: Advanced and Sustainable Development*, eds. Ahmed, S., Annu, S., and Ikram, S. Chapter 2 "Polymer materials: From the past to the future," (Nova Science Publishers, New York), pp. 35–52.

24. Arfin, T. (2018). *Handbook of Bionanocomposite: Green and Sustainable Materials*, eds. Ahmed, S. and Kanchi, S., Chapter 9 "MWCNT polymer composites: Environmental applications," (Jenny Stanford Publishing, Singapore), pp. 235–245.

25. Arfin, T, Tarannum, A., and Sonawane, K. (2018). *Green and Sustainable Advanced Materials: Processing and Characterization*, Vol. I, eds. Ahmed, S. and Hussain, C. M., Chapter 1 "Green and sustainable advanced materials: An overview," (Scrivener Publishing LLC, USA), pp. 1–34.

26. Arfin, T. and Tarannum, A. (2018). *Handbook of Nanomaterials for Industrial Applications,* ed. Hussain, C. M., Chapter 6 "Engineered nanomaterials for industrial application: An overview," (Elsevier, Netherlands), pp. 127–134.

27. Athar, S., Bushra, R., and Arfin, T. (2017). *Nanocellulose and Nanohydrogel Matrices: Biotechnological and Biomedical Applications*, eds. Jawaid, M. and Mohammad, Chapter 7 "Cellulose nanocrystals and PEO/PET hydrogel material in biotechnology and biomedicine: Current status and future prospects," (Wiley-VCH, Germany), pp. 139–173.

28. Borkar, R., Waghmare, S. S., and Arfin, T. (2017). *Nanocellulose and Nanohydrogel Matrices: Biotechnological and Biomedical Applications*, eds. Jawaid, M. and Mohammad, F., Chapter 2 "Bacterial cellulose and polyester hydrogel matrices in biotechnology and biomedicine: Current status and future prospects," (Wiley-VCH, Germany), pp. 21–46.

29. Khan, A. U., Malik, N., and Arfin, T. (2017). *Nanocellulose and Nanohydrogel Matrices: Biotechnological and Biomedical Applications*, eds. Jawaid, M. and Mohammad, F., Chapter 12 "Nanofibrillated cellulose and copoly (amino acid) hydrogel matrices in biotechnology and biomedicine," (Wiley-VCH, Germany), pp. 331–352.

30. Arfin, T. and Mohammad, F. (2016). Electrochemical, antimicrobial and anticancer effects of ethyl cellulose-nickel (II) hydrogen phosphate. *Innovations Corrosion Mater.* Sci., **6**, pp. 10–18.

31. Bender, H., Lehmann, J., and Wallenfels, K. (1959). Pullulan in extracellular glucans von *Pullularia pullulans. Biochim. Biophys. Acta,* **36**, pp. 309–316.

32. Bender, H. (1938). Physiology of *Dematium pullulans* de Bary. *Zentralbl. Bakteriol. Parasitenkd. Infektionskrankh. Hyg. Abt.*, **98**, pp. 133–167.

33. Cooke, W. B., Phaff, H. J., Miller, M. W., Shifrine, M., and Knapp, E. P. (1960). Yeasts in polluted water and sewage. *Mycologia,* **52**, pp. 210–230.

34. Gostinac, C., Ohm, R. A., Kogej, T., Sonjak, S., Turk, M., Zajc, J., Zalar, P., Grube, M., Sun, H., Han, J., Sharma, A., Chiniquy, J., Ngan, C. Y., Lipzen, A., Barry, K., Grigoriev, IV., and Gunde-Cimerman, N. (2014). Genome sequencing of four *Aureobasidium pullulans* varieties: Biotechnological potential, stress tolerance, and description of new species. *BMC Genom.,* **15**, pp. 549.

35. Prakitchaiwattana, C. J., Fleet, G. H., and Heard, G. M. (2004). Application and evaluation of denaturing gradient gel electrophoresis to analyse the yeast ecology of wine grapes. *FEMS Yeast Res.,* **4**, pp. 865–877.

36. Barata, A., Malfeito-Ferreira, M., and Loureiro, V. (2012). The microbial ecology of wine grape berries. *Int. J. Food Microbiol.,* **153**, pp. 243–259.

37. Suneetha, V., Sindhuja, K. V., and Sanjeev, K. (2010). Screening, characterization and optimization of pullulan producing microorganisms from plant leaves in Chittoor district. *Asian J. Microbiol. Biotechnol. Environ. Sci.,* **12**, pp. 149–155.

38. Carolan, G., Catley, B. J., and McDougal, F. G. (1983). The location of tetrasaccharide units in pullulan. *Carbohydr. Res.,* **114**, pp. 237–243.

39. Mishra, B., Vuppu, S., and Rath, K. (2011). The role of microbial pullulan, a biopolymer in pharmaceutical approaches: A review. *J. Appl. Pharm. Sci.,* **1**, pp. 45–50.

40. Simon, L., Caye-Vaugien, C., and Bouchonneau, M. (1993). Relation between pullulan production, morphological state and growth conditions in *Aureobasidium pullulans*: New observations. *J. Gen. Microbiol.,* **139**, pp. 979–985.

41. Duan, X. H., Chi, Z. M., Wang, L., and Wang, X. H. (2008). Influence of different sugars on pullulan production and activities of α-phospholglucose mutase, UDPG-pyrophosphorylase and glucosyltransferase involved in pullulan synthesis in *Aureobasidium pullulans* Y 68. *Carbohydr. Polym.,* **73**, pp. 587–593.

42. Ono, K., Kawahara, Y., and Ueda, S. (1977). Effect of H on pullulan elaboration by *Aureobasidium pullulans* S-1. *Agric. Biol. Chem.,* **44**, pp. 2113–2118.

43. Cheng, K. C., Demirci, A., and Catchmark, J. M. (2011). Continuous pullulan fermentation in a biofilm reactor, *Appl. Microbiol. Biotechnol.,* **90**, pp. 115–122.

44. Sugumaran, K. R. and Ponnusami, V. (2017). Review on production, downstream processing and characterization of microbial pullulan. *Carbohydr. Polym.*, **173**, pp. 573–591.

45. Cooke, W. B. (1959). An ecological life history of *Aureobasidium pullulans* (de Bary) Arnaud. *Mycopathol. Mycol. Appl.*, **12**, pp. 1–45.

46. Manitchotpisit, P., Skory, C. D., Leathers, T. D., Lotrakul, P., Eveleigh, D. E., Prasongsuk, S., and Punnapayak, H. (2011). Alpha-amylase activity during pullulan production and alpha-amylase gene analyses of *Aureobasidium pullulans*. *J. Ind. Microbiol. Biotechnol.*, **38**, pp. 1211–1218.

47. Leite, R. S., Bocchini, D. A., Martins Eda, S., Silva, D., Gomes, E., and Da Silva, R. (2007). Production of cellulolytic and hemicellulolytic enzymes from *Aureobasidium pullulans* on solid state fermentation. *Appl. Biochem. Biotechnol.*, **137–140**, pp. 281–288.

48. Leathers, T. D., Rich, J. O., Anderson, A. M., and Manitchotpisit, P. (2013). Lipase production by diverse phylogenetic clades of *Aureobasidium pullulans*. *Biotechnol. Lett.*, **35**, pp. 1701–1706.

49. Manitchotpisit, P., Leathers, T. D., Peterson, S. W., Kurtzman, C. P., Li, X. L., Eveleigh, D. E., Lotrakul, P., Prasongsuk, S., Dunlap, C. A., Vermillion, K. E., and Punnapayak, H. (2009). Multilocus phylogenetic analyses, pullulan production and xylanase activity of tropical isolates of *Aureobasidium pullulans*. *Mycol. Res.*, **113**, pp. 1107–1120.

50. Ni, X., Chi, Z., Ma, C., and Madzak, C. (2008). Cloning, characterization, and expression of the gene encoding alkaline protease in the marine yeast *Aureobasidium pullulans* 10. *Mar. Biotechnol.*, **10**, pp. 319–327.

51. Rich, J. O., Leathers, T. D., Anderson, A. M., Bischoff, K. M., and Manitchotpisit, P. (2013). Laccases from *Aureobasidium pullulans*. *Enzym. Microb. Technol.*, **53**, pp. 33–37.

52. Chi, Z., Wang, F., Chi, Z., Yue, L., Liu, G., and Zhang, T. (2009). Bioproducts from *Aureobasidium pullulans*, a biotechnologically important yeast. *Appl. Microbiol. Biotechnol.*, **82**, pp. 793–804.

53. Baffi, M. A., Tobal, T., Lago, J. H., Boscolo, M., Gomes, E., and Da-Silva, R. (2013). Wine aroma improvement using a beta-glucosidase preparation from *Aureobasidium pullulans*. *Appl. Biochem. Biotechnol.*, **169**, pp. 493–501.

54. Seo, H. P., Son, C. W., Chung, C. H., Jung, D. I., Kim, S. K., Gross, R. A., Kaplan, D. L., and Lee, J. W. (2004). Production of high molecular weight pullulan by *Aureobasidium pullulans* HP-2001 with soybean pomace as a nitrogen source. *Bioresour. Technol.*, **95**, pp. 293–299.

55. Filonow, A. B., Vishniac, H. S., Anderson, J. A., and Janisiewicz, W. J. (1996). Biological control of *Botrytis cinerea* in apple by yeasts from various habitats and their putative mechanisms of antagonism. *Biol. Control*, **7**, pp. 212–220.

56. Prajapati, V. D., Jani, G. K., and Khanda, S. M. (2013). Pullulan: An exopolysaccharide and its various applications. *Carbohydr. Polym.*, **95**, pp. 540–549.

57. Li, Y., Chi, Z., Wang, G. Y., Wang, Z. P., Liu, G. L., Lee, C. F., Ma, Z. C., and Chi, Z. M. (2015). Taxonomy of *Aureobasidium* spp. and biosynthesis and regulation of their extracellular polymers. *Crit. Rev. Microbiol.*, **41**, pp. 228–237.

58. Leathers, T. D. and Manitchotpisit, P. (2013). Production of poly(beta-L-malic acid) (PMA) from agricultural biomass substrates by *Aureobasidium pullulans*. *Biotechnol. Lett.*, **35**, pp. 83–89.

59. Kurosawa, T., Sakai, K., Nakahara, T., Oshima, Y., and Tabuch, T. (1994). Extracellular accumulation of the polyol lipids, 3,5-dihydroxydecanoyl and 5-hydroxy-2-decenoyl esters of arabitol and mannitol, by *Aureobasidium* sp. *Biosci. Biotechnol. Biochem.*, **58**, pp. 2057–2060.

60. Manitchotpisit, P., Watanapokasin, R., Price, N. P., Bischoff, K. M., Tayeh, M., Teeraworawit, S., Kriwong, S., and Leathers, T. D. (2014). *Aureobasidium pullulans* as a source of liamocins (heavy oils) with anticancer activity. *World J. Microbiol. Biotechnol.*, **30**, pp. 2199–2204.

61. Chi, Z., Wang, X. X., Ma, Z. C., Buzdar, M. A., and Chi, Z. M. (2012). The unique role of siderophore in marine-derived *Aureobasidium pullulans* HN6.2. *Biometals*, **25**, pp. 219–230.

62. Lin, Y., Zhang, Z., and Thibault, J. (2007). *Aureobasidium pullulans* batch cultivations based on a factorial design for improving the production and molecular weight of exopolysaccharides. *Process Biochem.*, **42**, pp. 820–827.

63. Bae, H., Ahari, A. F., Shin, H., Nichol, J. W., Hutson, C. B., Masaeli, M., Kim, S.-H., Aubin, H., Yamanlar, S., and Khademhosseini, A. (2011). Cell-laden microengineered pullulan methacrylate hydrogels promote cell proliferation and 3D cluster formation. *Soft Matter* **7**, pp. 1903–1911.

64. Campos, C. A., Gerschenson, L. N., and Flores, S. K. (2011). Development of edible films and coatings with antimicrobial activity. *Food Bioproc. Tech.*, **4**, pp. 849–875.

65. Barata, A., Malfeito-Ferreira, M., and Loureiro, V. (2012). The microbial ecology of wine grape berries. *Int. J. Food Microbiol.*, **153**, pp. 243–259.

66. Verginer, M., Leitner, E., and Berg, G. (2010). Production of volatile metabolites by grape-associated microorganisms. *J. Agric. Food Chem.*, **58**, pp. 8344–8350.

67. Campisano, A., Pancher, M., Puopolo, G., Puddu, A., Lòpez-Fernàndez, S., Biagini, B., Yousaf, S., and Pertot, I. (2014). Diversity in endophytic populations reveals functional and taxonomic diversity between wild and domesticated grapevines. *Am. J. Enol. Vitic.*, **66**, pp. 12–21.

68. Leathers, T. D. (2003). Biotechnological production and applications of pullulan. *Appl. Microbiol. Biotechnol.*, **62**, pp. 468–473.

69. Singh, R. S., Saini, G. K., and Kennedy, J. F. (2008). Pullulan: Microbial sources, production and applications. *Carbohydr. Polym.*, **73**, pp. 515–531.

70. Ürküt, Z., Dagbaglı, S., and Goksungur, Y. (2007). Optimization of pullulan production using-Ca-alginate immobilized *Aureobasidium pullulans* by response surface methodology. *J. Chem. Tech. Biotechnol.*, **82**, pp. 837–846.

71. Farris, S., Unalan, I. U., Introzzi, L., Fuentes-Alventosa, J. M., and Cozzolino, C. A. (2014). Pullulan-based films and coatings for food packaging: Present applications, emerging opportunities, and future challenges. *J. Appl. Polym. Sci.*, **131**, pp. 40539.

72. Krochta, J. M. and De Mulder-Johnston, C. (1997). Edible and biodegradable polymer films: Challenges and opportunities. *Food Technology*, **51**, pp. 61–74.

73. Singh, R. S. and Kaur, N. (2018). Biochemical and molecular characterization of a new pullulan producer *Rhodosporidium paludigenum* PUPY-06, *J. Appl. Biol. Biotechnol.*, **6**, pp. 28–37.

74. Carrington, A. K., Goff, H. D., and Stanley, D. W. (1996). Structure and stability of the glassy state in rapidly and slowly cooled carbohydrate solutions. *Food Res. Int.*, **29**, pp. 207–213.

75. Gaidhani, H. K., McNeil, B., and Ni, X. (2005). Fermentation of pullulan using an oscillatory baffled fermenter. *Chem. Eng. Res. Des.*, **83**, pp. 640–645.

76. Oguzhan, P. and Yangilar, F. (2013). Pullulan: Production and usage in food industry. *Af. J. Food Sci. Technol.*, **4**, pp. 57–63.

77. Yim, H., Yang, S.-G., Jeon, Y. S., Park, I. S., Kim, M., Lee, D. H., Bae, Y. H., and Na, K. (2011). The performance of gadolinium diethylene triamine pentaacetate pullulan hepatocyte-specific T1 contrast agent for MRI. *Biomaterials*, **32**, pp. 5187–5194.

78. Ye, S., Ma, Z., Liu, Z., Liu, Y., Zhang, M., and Wang, J. (2014). Effects of carbohydrate sources on biosorption properties of the novel exopolysaccharides produced by arthrobacter ps-5. *Carbohydr. Polym.*, **112**, pp. 615–621.

79. Li, W. W., Zhou, W. Z., Zhang, Y. Z., Wang, J., and Zhu, X. B. (2008). Flocculation behavior and mechanism of an exopolysaccharide from the deep-sea psychrophilic bacterium *Pseudoalteromonas* sp. SM9913. *Bioresour. Technol.*, **99**, pp. 6893–6899.

80. Constantin, M., Asmarandei, I., Harabagiu, V., Ghimici, L., Ascenzi, P., and Fundueanu, G. (2013). Removal of anionic dyes from aqueous solutions by an ion-exchanger based on pullulan microspheres. *Carbohydr Polym.*, **91**, pp. 74–84.

81. Ghimici, L. and Constantin, M. (2015). The separation of the pyrethroid insecticide fastac 10 EC by cationic pullulan derivatives. *React. Funct. Polym.*, **95**, pp. 12–18.

82. Arfin, T. and Mohammad, F. (2015). *Advances in Materials Science Research*, Vol. 21, ed. Wythers, M. C., Chapter 7 "Dendrimer and its role for the advancement of nanotechnology and bioengineering," (Nova Science Publishers, New York), pp. 157–174.

83. Arfin, T., Mohammad, F., and Yusof, N. A. (2015). *Polystyrene: Synthesis, Characteristics and Application*, ed. Lynwood, C., Chapter 10 "Applications of polystyrene and its role as a base in industrial chemistry," (Nova Science Publishers, New York), pp. 269–280.

84. Arfin, T., Varshney, N., and Singh, B. (2019). *Green Materials for Wastewater Treatment. Environmental Chemistry*, Vol, 38, eds. Naushad, M. and Lichtfise, E., Chapter 11 "Ionic liquid modified activated carbon for the treatment of textile wastewater," (Springer, Cham), pp. 257–275.

85. Onwudiwe, D. C., Arfin, T., and Strydom, C. A. (2014). Synthesis, characterization, and dielectric properties of N-butyl aniline capped CdS nanoparticles. *Electrochim. Acta*, **116**, pp. 217–223.

86. Arfin, T. (2018). *Handbook of Bionanocomposite: Green and Sustainable Materials*, eds. Ahmed, S. and Kanchi, S., Chapter 9 "MWCNT polymer composites: Environmental applications," (Jenny Stanford Publishing, Singapore), pp. 235–245.

87. Sophia, A. C., Arfin, T., and Lima, E. C. (2019). *A New Generation Material Graphene: Applications in Water Technology*, ed. Naushad, M., Chapter 18 "Recent developments in adsorption of dyes using graphene based nanomaterials," (Springer, Cham), pp. 439–471.

88. Arfin, T., Sonawane, K., Saidankar, P., and Sharma, S. (2019). *Integrating Green Chemistry and Sustainable Engineering*, ed. Shahid-ul-Islam, Chapter 14 "Role of microbes in the bioremediation of toxic dyes," (Scrivener Publishing LLC, MA), pp. 443–472.

89. Arfin, T. and Sonawane, K. (2019). *Integrating Green Chemistry and Sustainable Engineering*, ed. Shahid-ul-Islam, Chapter 18 "Biotechnology past-to-future," (Scrivener Publishing LLC, MA), pp. 617–645.

90. Mohammad, F. and Arfin, T. (2014). Thermodynamics and electrochemical characterization of core-shell type gold-coated superparamagnetic iron oxide nanoparticles, *Adv. Mater. Lett.*, **5**, pp. 315–324.

91. Waghmare, S., Arfin, T., Manwar, N., Lataye, D., Labhsetwar, N., and Rayalu, S. (2015). Preparation and characterization of *Polyalthia longifolia* based adsorbent for removing fluoride from drinking water, *Asian J. Adv. Basic Sci.*, **4**, pp. 12–24.

Chapter 3

Synthesis of Pullulan

Neetu Singh Tanwar
Department of Chemistry, YMCA University of Science and Technology, Sector-6, Faridabad 121006, Haryana, India
1980neetu@gmail.com

This chapter describes the process of synthesizing pullulan and the various factors on which its purity and yield depend. Pullulan is formed extracellularly when the cells are in the late log phase and stationary phase of growth. Its formation depends on various factors, including pH, temperature, pressure, substrate, carbon/nitrogen ratio, concentration of dissolved oxygen, fermentation time and strain. The yield and molecular weight of pullulan can be adjusted by manipulating the substrate and fermentation conditions. *Aureobasidium pullulans* (a polymorphic fungus) plays an important role as a precursor in the synthesis of pullulan.

3.1 Introduction

Pullulan is an extracellular, linear homopolysaccharide consisting of maltotriose and maltotetraose with α-(1→6) and α-(1→4) linkages.

Pullulan: Processing, Properties, and Applications
Edited by Shakeel Ahmed and Aisverya Soundararajan
Copyright © 2021 Jenny Stanford Publishing Pte. Ltd.
ISBN 978-981-4877-17-6 (Hardcover), 978-1-003-05694-2 (eBook)
www.jennystanford.com

The molecular weight of pullulan having the molecular formula $(C_6H_{10}O_5)_n$ is in the range of 10–3000 kDa [1, 12, 14, 21]. This unique structure makes pullulan highly soluble in water and flexible. Pullulan is nontoxic, nonmutagenic, noncarcinogenic, biodegradable, and edible [9, 24]. Pullulan has been successfully synthesized for a few decades under certain conditions such as time duration, pH, and phosphate concentration from the culture broth/(fungal strain) of *Aureobasidium pullulans* (for example *Aureobasidium pullulans, Tremella mesenterica, Cytaria harioti, Cytaria darwinii, Cryphonectria parasitica, Teloschites flavicans, Rhodototula bacarum*, and *Aureobasidium pullulans*). The synthesis of other polysaccharide varieties, similar to pullulan, has been done by other microbes [2, 22, 25, 28]. It has been well established that the regularly repeating unit of pullulan is a maltotriose trimer (produced extracellularly by *A. pullulans*) α-(1→4)Glu*p*- α-(1→4)Glu*p*- α-(1→6)Glu*p*-. However, other repeating units in pullulan, tetramer or maltotetraose α-(1→4)Glu*p*- α-(1→4)Glu*p*- α-(1→4)Glu*p*- α-(1→6)Glu*p*-, may be present in the pullulan polymeric chain. But the maximum extent of maltotetraose subunits has been found to be only 7% [7].

3.2 Synthesis of Pullulan

Pullulan is synthesized by *A. pullulans* (formally known as pullan pullularia). The organism (*A. pullulans*) involved is ubiquitous in nature and is found in soil, lake water, on the surface of latex paint files, synthetic plastic materials, used cosmetic, and foods such as cereals, fruits, cheese, tomato, etc. This organism forms a black pigment (melanin), so it is also called "black yeast." Different synthesis routes for the preparation of pullulan have been proposed by several researchers [6, 19]. In 1981, Catley reported that up to 7% maltotetraose subunits were detected; this is the main reason why, currently and frequently in the literature, the term "pullulan" is used for both the "polymaltotriose" produced by *A. pullulans* and the polysaccharide varieties. Catley proposed that a lipid is used as a carrier to take out pullulan outside of plasmalemma. Rho et al. proposed two pathways; one allowed direct conversion of glucose, whereas in the second pathway, an identified precursor is used [26].

In the production process of pure pullulan, different routes are taken such as filtration, decoloration, removal of extra salt and proteins, and finely drying and crushing.

3.2.1 Role of *A. pullulans* in the Synthesis of Pullulan

For the production of pullulan on industrial scale, mesophilic fermentation of liquefied (hydrolyzed) starch is carried out under controlled conditions, and a strain of *A. pullulans* specifically chosen should not be genetically modified, pathogenic, and toxigenic [12]. Pullulan is produced by a yeast-like fungus *P. pullulans*

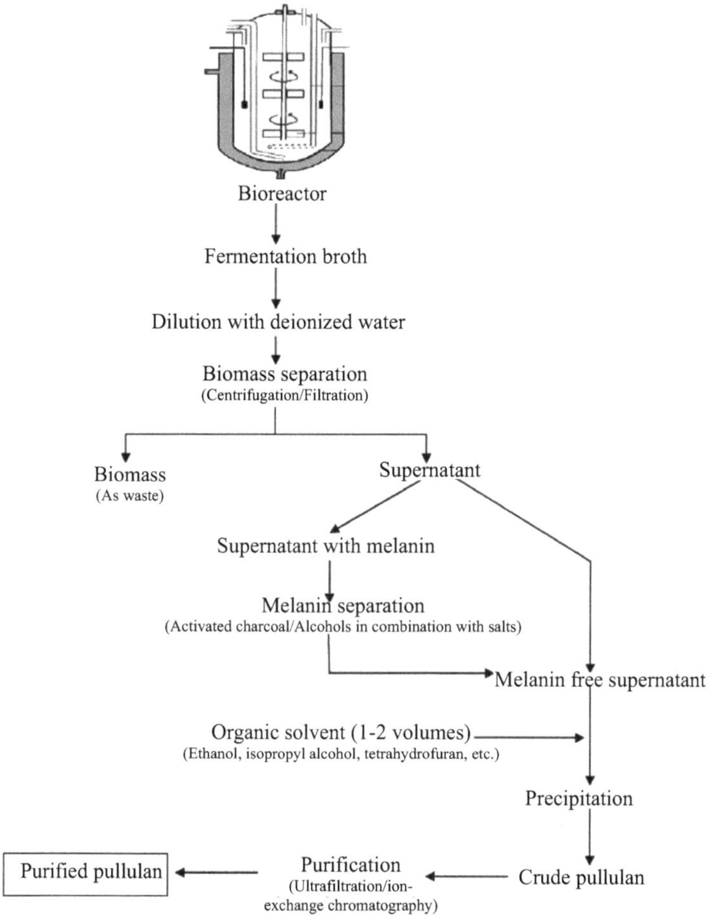

Figure 3.1 Flow diagram of downstream processing of pullulan.

from glycosidic carbon sources and also non-glycosidic substrates (glycerol and acetate). The main impurities are mono-, di-, and oligosaccharides, which may come from raw material (hydrolyzed starch) in the final product.

A. pullulans is a polymorphic fungus, and it has three distinctive forms: elongated ranched septate (also called septa or cross-walls, which divide the hyphae into various uninucleated or multinucleated cells) filaments, large chlamydospores, and smaller elliptical yeast-like cells. The two main undesirable characteristic features of *A. pullulans* are production of a dark pigment (melanin-like compound, which appears dark green to black in color) and second occur during its submerged growth. As the fermentation process goes on, the viscosity of the culture medium decreases because of the decrease in the average molecular weight of the accumulated extracellular pullulan [5, 17]. Pullulan formation is accomplished with the morphology shift from mycelia to unicellular forms, and consequently blastospores play the role to elaborate pullulan [11, 27]. Some researchers also suggest that swollen cells and chlamydospores are also pullulan producers [29]. Pullulan can be produced by a number of (but not all) strains of *A. pullulans* [3, 20, 32]. Other than *A. pullulans*, a number of other microorganisms are also reported for pullulan production [8, 10, 23, 33]. The pullulan recovered is characterized by ^{13}C NMR spectroscopy, which confirms the sequence of both maltotriose and maltotetraose in the polysaccharide backbone [13].

3.2.2 Isolation of *A. pullulans* from Different Sources to Synthesize Pullulan

Several undesired features have been seen in the synthesis process of pullulan by *A. pullulans*, for example formation of intracellular and extracellular black pigments during the fermentation of *A. pullulans*, which also comes in the final product; decrease in the molecular weight of the accumulated extracellular pullulan; and also decrease in yield. The yield of pullulan depends on various factors, such as type of *A. pullulans*, pH, temperature, culture viscosity, etc. So the selection of *A. pullulans* is a very important task. Numerous studies have been done in this context, including isolation of *A. pullulans* from various sources [30].

Pollock et al. reported strain designations, synonyms, and sources for preexisting and isolates of *A. pullulans* and *Saccharomyces cerevisiae* (listed in Table 3.1).

Table 3.1 Different fungal strains

No.	Original name	Source
F6	*S. cerevisiae* DL1α	*S. subramani*, UCSD
API	IF04464	Fermentation Institute, Osaka, Japan
AP2	IF06353	Fermentation Institute, Osaka, Japan
AP3	QM3092	ATCC 9348
AP4	NRRL Y6220	ATCC 34647
AP5	NRRL Y12996	ATCC 42023
AP7	S-1	NRRL Y17005
AP9	Y12974	NRRL, Peoria, Ill

*All strains, except DL1α, are *A. pullulans.*

In the synthesis of pullulan, both carbon and nitrogen sources show their importance. The carbon sources studied include sucrose, glucose, fructose, maltose, and corn syrup, while the nitrogen sources studied are ammonium nitrate, ammonium sulfate, ammonium tartrate, sodium nitrate, asparagine, and urea. The choice of C and N sources and their concentration very significantly affect the production of pullulan.

3.2.2.1 Isolation process of *A. pullulans*

Fungal strains and culture media: Shake flask culture can be grown in either a complex nitrogen medium (P1) or a minimal salt medium (P2). Medium P1 includes: yeast extract, NH_4SO_4, NaCl, $MgSO_4$, $ZnSO_4$, and HCl to pH 6.0. Medium P2 includes: $(NH_4)_2HPO_4$, NaCl, $MgSO_4 \cdot 7H_2O$, K_2HPO_4, $FeSO_4$, $MnSO_4$, $ZnSO_4$, and HCl to pH 7.0. A carbon source (e.g., corn) syrup is being added to (1% to 5% gram per 100 ml) and agar includes at 15 g/l for solid phase.

Isolation of *A. pullulans* is done by the selective enrichment method as followed by Pollock et al. [30]. They did work on plant leaf surfaces to produce *A. pullulans.* The new fungal isolates were identified as *A. pullulans* on the basis of the appearance of polymorphic colonies formed on agar plates, the electrophoretic profiles of repeated genomic DNA sequences, and the production of pullulan in shake flask cultures. In this process, a leaf is detached from each plant and soaked in sterile water for 3 days at 25°C, and then 0.1 ml is transferred to 10 ml of P2 minimal salts medium (pH 4) containing 1% (wt/vol) carbon source syrup and 10 μg of chloramphenicol per ml. After 2 days of shaking at 25°C, the turbid culture is allowed to rest undisturbed for 20 min to allow filaments and aggregates to settle to the bottom. About 10 μl from the upper, partially clarified phase that was enriched for yeast-like cells was spread onto agar plates containing P2 medium (pH 5), sucrose syrup, and 10 μg of chlorophenicol per ml. After 4 days, independent colonies are purified by replating and tested in shake flasks for the secretion of isopropyl alcohol (IPA)-precipitable polysaccharides and culture pigmentation. Different isolates showed different degrees of pigmentations.

Melanin pigment removal from the fermentation broth can be done by hydrogen peroxide treatment [4]. Different levels of hydrogen peroxide are added to the supernatant collected from the centrifuged broth at room temperature and allowed to stir slowly for 15–20 min to determine the optimum amount of hydrogen peroxide required for the oxidation process with the removal of the melanin pigments. The OD at 320 nm was checked for each concentration of the utilized H_2O_2 in order to find the amount of melanin pigment.

3.2.3 Recovery of Pullulans

The recovery of pullulan is carried out first by centrifuging the cells and then keeping the resulted supernatant at 4°C, following the addition of twice volumes of cold isopropyl alcohol. The residual hydrogen peroxide, which was utilized for the removal of melanin from the fermentation broth, is discarded with the supernatant after the precipitation of pullulan. The precipitate is washed twice with distilled water and dried at 80°C till a constant weight. The separation of biomass by centrifugation is done before the addition of isopropanol and taken as pellet. The pellet is placed in an aluminum foil and dried overnight at 80°C. The recovered pullulan is estimated as grams of pullulan (dry weight) produced per x ml of fermented broth.

3.3 Factors Affecting Production of Pullulan

3.3.1 Influence of Concentration of Dissolved Oxygen on Production of Pullulan

Dissolved oxygen has a significant role in pullulan fermentation. It has been proved that the effect of oxygen on aerobic microbial processes is quite different and varies from species to species [1]. The role of oxygen for pullulan production is different for nitrogen-containing medium and nitrogen-free medium [26]. In a nitrogen medium, the yield and production rate of pullulan increase with oxygen availability, while in a nitrogen-free medium, oxygen availability shows inverse effect. Pullulan production rate is also influenced by stirrer speed [20]. The rotating rate controls the level of dissolved oxygen. The yield of pullulan is specific for an influx rate of oxygen. Under anaerobic conditions, fermentation is completely ceased. The availability of oxygen for fermentation also depends on strain ploidy [26].

3.3.2 Effect of pH and Pullulan Synthesis

The morphology of an organism that influences pullulan synthesis is significantly affected by the initial pH of fermentation medium [15]. Optimum conditions are to be set for higher production of pullulan. HCl or NaOH is used to adjust the pH. Different studies suggest that a pH within the range of 4 to 7 is suitable for pullulan synthesis. It can be understood by an example in which the effect of initial pH (4.0 to 8.0) on the kinetics of pullulan from *A. pullulans* utilizing coconut water and coconut milk extract is shown in Fig. 3.2. Pullulan concentration gradually increased with increasing initial pH up to 7 and then decreased. The relative concentrations of pullulan at pH 4 and 8 were 10.7 g/L and 30.5 g/L, respectively. The highest pullulan concentration of 58.0 g/L was achieved at pH of 7.0. A lower production of cell mass and pullulan was obtained at lower pH levels when compared to higher levels of pH. This is probably due to an influence of acidic pH on the morphological character of the organism [16, 18].

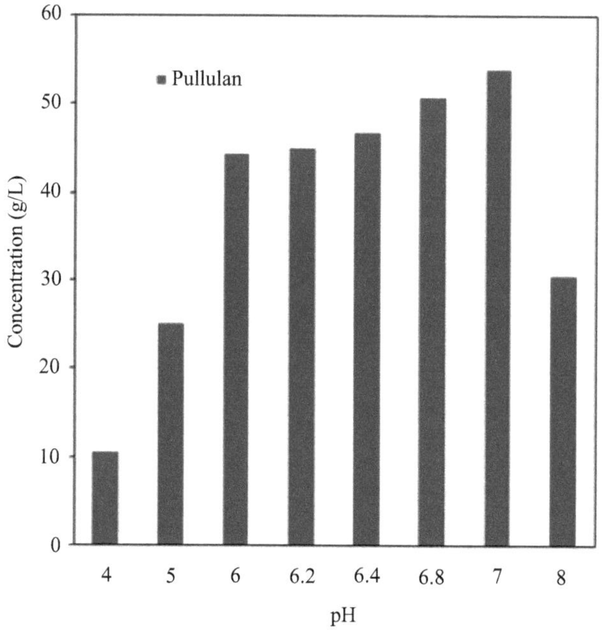

Figure 3.2 Effect of pH on pullulan production using coconut milk by *Aureobasidium pullulans*.

3.3.3 Effect of Fermentation Time on Biosynthesis of Pullulan

In order to find an optimum fermentation time for pullulan fermentation, experiments are carried out for different fermentation times. The kinetics of pullulan synthesis is studied, and it is seen that pullulan concentration gradually increases with fermentation time and reaches a maximum for a certain fermentation period. After which, the production starts to decrease; it may be because of the hydrolysis of pullulan by endogenous-A, released by the organism at the latter stages of fermentation [31]. But several other factors are also responsible for such variability of the pullulan yield, including organism used, chemical composition of substrate, fermentation system, and conditions provided during the fermentation [16, 31].

3.3.4 Effect of Nitrogen Source on Pullulan Synthesis

Different nitrogen sources such as $(NH_4)_2SO_4$, NH_4Cl, $NaNO_3$, $NaNO_2$, malt extract, peptone, and yeast extract are studied to attain maximum production of pullulan. The nitrogen source equivalent is kept constant for all nitrogen sources in all experiments [5].

Depletion of nitrogen is must for the initiation of pullulan synthesis. It has been seen that 10:1 C/N ratio is more favorable for pullulan synthesis [16].

3.4 Conclusion

Research studies have proved that pullulan is an extracellular water-soluble microbial polysaccharide produced by strains of *A. pullulans.* It is a linear α-D-glucan connected with α-1,4 glycosidic bond mainly of maltotriose repeating units interconnected by α-1,6 linkages. The regular alternation of α-1,4 and α-1,6 bond results in two distinctive properties, the structural flexibility and enhanced solubility. Researchers have shown deep interest in pullulan as it is nontoxic, nonmutagenic, noncarcinogenic, biodegradable, and edible. *A. pullulans* is a polymorphic fungus that synthesizes several polysaccharides, including pullulan. Most of the studies reveal that mesophilic strains of this fungus produce pullulan in flask shake experiments or in fermenter experiments under certain physiochemical conditions. Non-stirred fermentation system is beneficial over stirred fermentation system because it cuts down the cost and energy. Therefore, strains capable of producing high amount of pullulan in non-stirred fermentation system may be beneficial for commercial production of pullulan. Dissolved oxygen concentration affects pullulan fermentation in different ways for nitrogen-containing and nitrogen-free mediums. Pullulan synthesis is also affected by the pH of fermentation medium. It is found that 4 to 7 pH range is most suitable for pullulan production. Similarly, nitrogen sources are also chosen to attain maximum pullulan yield and 10:1 C/N ratio shows good results in this perspective.

References

1. Wecker, A. and Onken, U. (1991). Influence of dissolved oxygen concentration and shear rate on the production of pullulan by *Aureobasidium pullulans*, *Biotechnol. Lett.*, **13**, pp. 155–160A.
2. Auer, D. P. F. and Seviour, R. J. (1990). Influence of varying nitrogen sources on polysaccharide production by *Aureobasidium pullulans* in batch culture, *Appl. Microbiol. Biotechnol.*, **32**, pp. 637–644.
3. Augustin, J., Kuniak, L., and Hudecova, D. (1997). Screening of yeasts and yeast-like organisms *Aureobasidium pullulans* for pullulan production, *Biologia Bratislava*, **52**, pp. 399–404.
4. Mishra, B. and Vuppu, S. (2013). A study on downstream processing for the production of pullulan by *Aureobasidium pullulans*-SB-01 from the fermentation broth, *Res. J. Recent Sci.*, **2**, pp. 16–19.
5. Catley, B. J. (1971a). Role of pH and nitrogen limitation in the elaboration of the extracellular polysaccharide pullulan by *Pullularia pullulans*, *Appl. Microbiol.*, **22**, pp. 650–654.
6. Catley, B. J. (1971b). Utilization of carbon sources by *Pullularia pullulans* for the elaboration of extracellular polysaccharides, *Appl. Microbiol.*, **22**, pp. 641–649.
7. Catley, B. J., Ramsay, A., and Servis, C. (1986). Observations on the structure of the fungal extracellular polysaccharide, pullulan, *Carbohydr. Res.*, **153**, pp. 79–86.
8. Corsaro, M. M., De Castro, C., Evidente, A., Lanzetta, R., Molinaro, A., Parilli, M., and Sparapano, L. (1998). Phytotoxic extracellular polysaccharide fractions from *Cryphonectria parasitica* (Murr) Barr strains, *Carbohydr. Polym.*, **37**, pp. 167–172.
9. Kumar, D., Saini, N., Pandit, V., and Ali, S. (2012). An insight to pullulan: A biopolymer in pharmaceutical approaches, *Int. J. Basic Appl. Sci.*, **1**(3), pp. 202–219.
10. Fraser, C. G. and Jennings, H. J. (1971). A glucan from *Tremella mesenterica* NRRL Y6158, *Can. J. Chem.*, **49**, pp. 1804–1807.
11. Gibbs, P. A. and Seviour, R. J. (1996). Pullulan. In: *Polysaccharides in Medicinal Applications* (S. Dimitiu, Ed.), pp. 59–86. New York: Marcel Dekker.
12. Moubashera, H. and Wahshb, S. S. (2014). Pullulan production from *Aureobasidium pullulans* by continuous culture, *Basic Res. J. Microbiol.*, **1**(6), pp. 11–15.

13. Jennings, H. J. and Smith, I. C. P. (1973). Determination of the composition and sequence of a glucan containing mixed linkages by carbon-13 nuclear magnetic resonance, *J. Am. Chem. Soc.*, **92**, pp. 606–608.

14. Lee, J. W., Yeomans, W. G., Allen, A. L., Deng., F., Gross, R. A., and Kaplan, D. L. (1999). Biosynthesis of novel exopolymers by *Aureobasidium pullulans*, *Appl. Environ. Microbiol.*, **65**(12), pp. 5265–5271.

15. Thriumavalavan, K., Manikkandan, T. R., and Dhanasekar, R. (2008). Batch fermentation kinetics of pullulan from *Aureobasidium pullulans* using low cost substrates, *Biotechnology*, **7**, pp. 317–322.

16. Thirumavalavan, K., Manikkadan, T. R., and Dhanasekar, R. (2009). Pullulan production from coconut by-products by *Aureobasidium pullulans*, *Afr. J. Biotechnol.*, **8**, pp. 254–258.

17. Kaplan, D. L., Wiely, B. J., Arcidiacono, S., Mayer, J., and Sousa, S. (1987). Controlling biopolymer molecular weight distribution: Pullulan and chitosan. In: *First Material Biotechnology Symposium* (D. L. Kaplan, Ed.), pp. 149–173. Technical Report NATICK/TR-88/033. Natick, MA; US Army.

18. Lacroix, C., LeDuy, A., Noel, G., and Choplin, L. (1985). Effect of pH on the batch fermentation of pullulan from sucrose medium, *Biotechnol. Bioeng.*, **27**, pp. 202–207.

19. Lee, J. W., Yeomans, W. G., Allen, A. L., Deng, F., Gross, R. A., and Kaplan, D. L. (1999). Biosynthesis of novel exopolymers by *Aureobasidium pullulans*, *Appl. Environ. Microbiol.*, **65**, pp. 5265–5271.

20. Leathers, T. D. (1987). Host amylases and pullulan production. In: *First Materials Biotechnology Symposium* (D. L. Kaplan, Ed.), pp. 175–185. Technical report NATICK/TR-88/033. Natick, MA: US Army.

21. McNeil, B. and Kristiansen, B. (1990). Temperature effects on polysaccharide formation by *Aureobasidium pullulans* in stirred tanks, *Enzyme Microb. Technol.*, **12**, pp. 521–526.

22. Ono, K., Yasuda, N., and Ueda, S. (1977). Effect of pH on pullulan elaboration by *Aureobasidium pullulans,* S-1, *Agric. Biol. Chem.*, **41**, pp. 2113–2118.

23. Oliva, E. M., Cirelli, A. F., and De Lederkremer, R. M. (1986). Characterization of a pullulan in *Cyttaria darwinii. Carbohydr. Res.*, **158**, pp. 262–267.

24. Oğuzhan, P. and Yangılar, F. (2013). Production and usage in food industry, *Afr. J. Food Sci. Technol.*, **4**(3), pp. 57–63.

25. Singh, R. S., Saini, G. K., and Kennedy, J. F. (2008). Pullulan: Microbial sources, production and applications, *Carbohydr. Polym.*, **73**, pp. 515–531.
26. Rho, D., Mulchandani, A., Luong, J. H. T., and LeDuy, A. (1988). Oxygen requirement in pullulan fermentation, *Appl. Microbiol. Biotechnol.*, **2**, pp. 361–366.
27. Reeslev, M., Storm, T., Jensen, B., and Olsen, J. (1997). The ability of yeast form of *Aureobasidium pullulans* to elaborate exopolysaccharide in chemostat culture at various pH values, *Mycol. Res.*, **101**, pp. 650–652.
28. Seviour, R. J. and Kristiansen, B. (1983). Effect of ammonium ion concentration on polysaccharide production by *Aureobasidium pullulans* in batch culture, *Eur. J. Appl. Microbiol. Biotechnol.*, **17**, pp. 178–181.
29. Simon, I., Bouchet, B., Caye-Vaugien, C., and Gallant, D. J. (1995). Pullulan elaboration and differentiation of the resting forms in *Aureobasidium pullulans, Can. J. Microbiol.*, **41**, pp. 35–45.
30. Pollock, T. J., Thorne, L., and Armentrout, R. W. (1992). Isolation of new *Aureobasidium* strains that produce high-molecular-weight pullulan with reduced pigmentation, *Appl. Environ. Microbiol.*, **58**(3), pp. 877–883.
31. West, T. P. and Strohfus, B. (1996). Pullulan production by *Aureobasidium pullulan* grown on ethanol stillage as a nitrogen source, *Microbios*, **88**, pp. 7–18.
32. Ueda, S., Fujita, K., Komatsu, K., and Nakashima, Z. (1963). Polysaccharide produced by the genus *Pullularia*. I. Production of polysaccharide by growing cells, *Appl. Microbiol.*, **11**, pp. 211–215.
33. Waksman, N., De Lederkremer, R. M., and Cerezo, A. S. (1977). The structure of an a-D-glucan from *Cyttaria harioti* Fischer. *Carbohydr. Res.*, **59**, pp. 505–515.

Chapter 4

Factors Affecting Pullulan Production

Ashwini Ravi,[a] S. Vijayanand,[a] G. Ramya,[a] Aisverya Soundararajan,[b] P. N. Sudha,[b] and J. Hemapriya[c]
[a]*Bioresource Technology Lab, Department of Biotechnology, Thiruvalluvar University, Serkkadu, Vellore 632 115, Tamil Nadu, India*
[b]*Biomaterial Research Lab, Department of Chemistry, DKM College for Women (Autonomous), Vellore 632 001, Tamil Nadu, India*
[c]*PG & Research Department of Microbiology, DKM College for Women (Autonomous), Vellore 632 001, Tamil Nadu, India*
micropriya80@gmail.com

Upstream processing is an important aspect of any production industries, especially industries dealing with microbial interaction. The process deals with many components such as selection of raw materials, processing raw materials, incubation, separation, purification, and getting a final product. Among these, when microbial process is considered, the incubation period makes highly remarkable changes in the raw materials to get the desired product. To achieve this perfect incubation period, several parameters such as carbon source, nitrogen source, temperature, pH, etc. have to be

Pullulan: Processing, Properties, and Applications
Edited by Shakeel Ahmed and Aisverya Soundararajan
Copyright © 2021 Jenny Stanford Publishing Pte. Ltd.
ISBN 978-981-4877-17-6 (Hardcover), 978-1-003-05694-2 (eBook)
www.jennystanford.com

optimized. Any minor changes in these parameters affect the final product either in quantity or in quality. Therefore, prior to large-scale production, small-scale testing for optimizing incubatory conditions is mandatory. Pullulan is an exopolysaccharide produced by yeast-like microfungus *Aureobasidium pullulans.* Pullulan was found to have several applications in the field of medicine, food, and environment. In order to explore other applications and to utilize them for the present applications, producing them in large quantities is amendable. To achieve the better production of pullulan, optimizing its incubation parameters is necessary. Therefore, in the present chapter, various parameters that cause production of pullulan in desirable amounts and the factors that affect their production are discussed in detail.

4.1 Introduction

Pullulan is an exopolysaccharide produced by the yeast-like fungus *A. pullulans* [5]. The chemical structure of these polysaccharides varies with substrate. Several studies on pullulan confirmed that it is a water-soluble neutral glucan consisting of either linear chains of glucopyranose units with regular alteration of two α (1$\rightarrow$4) and one (1$\rightarrow$6) linkages or a linear polymer of maltotriose units connected by (1$\rightarrow$6) linkages [93]. The structure of pullulan is given in Fig. 4.1. Apart from *A. pullulans*, pullulan is produced by various organisms such as *Tremella mesenterica* [19], *Cytaria harioti* [49, 79], *Cytaria darwinii* [49, 79], *Cryphonectria parasitica* [17], *Teloschites flavians* [54], and *Rhodototulo bacarum* [15].

The important physical properties of pullulan are its solubility in water and viscosity. Because of this, it is used in various commercial applications [73]. Since pullulan is biodegradable, oxygen impermeable, and edible, it is used in a wide variety of food products as non-polluting wrapping material for food supplements, binder, lubricant, gelling agent, emulsifier, and stabilizer [40, 69, 73]. In the pharmaceutical industry, pullulan is used as denture adhesive, oral care strips, drug delivery, gene delivery, vaccination, medical imaging, molecular chaperones, and film forming [11, 36–38, 59, 65, 72]. Due to its vast applications, pullulan is produced commercially.

A. pullulans and various modifications have been made in the culture medium to increase the efficiency of pullulan production. The various media supplements and changes made in them for better production of pullulan are discussed in the present chapter.

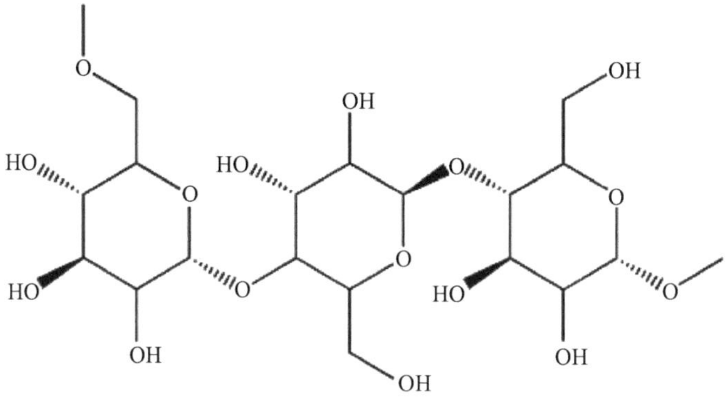

Figure 4.1 Structure of pullulan [28].

4.2 Medium for Pullulan Production

The first commercial production of pullulan was initiated in 1976 from hydrolyzed starch substrate by Hayashibara Company, Okayama, Japan. Later in 1982, pullulan films were produced by the same company [40, 76]. Production of pullulan is an aerobic process, which involves supply of oxygen to the liquid medium. In the commercial production of pullulan, starch hydrolysates of dextrose equivalents along with peptone, phosphate, and basal salts are used as medium. Production is performed in batch culture, at initial pH 6.5, which then falls to 3.5. The cultures are stirred continuously for providing aeration, and temperature is maintained at 30°C. The maximal culture growth is obtained at 75 H, and pullulan is produced at 100 H. Some species of *A. pullulans* that produce melanin are called "black yeast" [73]. As some strains of *A. pullulans* also produce melanin, the culture conditions and selection of strain are much important in obtaining pullulan of high molecular weight, which is relatively free of melanin [76].

Several studies on pullulan production from *A. pullulans* showed that the strain differs with growth, pullulan yield, and cell morphology. A wide variety of culture compounds are used in pullulan production, which are given in Table 4.1 [73].

Table 4.1 Culture components used for pullulan production

Source	Types	Reference
Carbon	Sucrose, glucose, fructose, maltose, maltooligosaccharides, xylose, arabinose, mannose, lactose, galactose, rhamnose	3, 4, 9
Nitrogen	Ammonium, yeast extract, peptone	52, 62, 76, 82
Minerals	Zn^{2+}, Fe^{2+}, Mn^{2+}, Ca^{2+}, Cu^{2+}	53
Vitamin	Biotin	83

The main problem with upstream processing of pullulan production is the cost of raw materials and optimization of process parameters [45, 48]. The cost of pullulan is estimated to be higher than the cost of other polysaccharides. This increased cost of pullulan is attributed to two factors [5, 25]:

1. The higher cost of raw materials
2. Lesser yield of pullulan since some strains cannot tolerate high levels of glucose in the culture medium

For selection of strains, research has been performed extensively, and strains such as *Rhodosporidium paludigenum, Rhodotorula bacarum*, and a mutant of *A. pullulans* have been found to produce pullulan efficiently [12, 70]. The second factor affecting upstream processing is the culture conditions of pullulan production. Statistical design and single optimization study have been performed to optimize culture conditions. In all the studies, the major drawback was associated with costly culture compounds such as yeast extract, peptone, and sucrose. Therefore, both drawbacks of pullulan production directly and indirectly point toward culture components. Many researchers around the world are interested in studying alternative sources for pullulan production [68].

4.3 Factors Affecting Pullulan Production

Pullulan has been produced by various researchers using different fermentation techniques such as batch fermentation [78], fed batch fermentation [46], and continuous fermentation [60]. Pullulan production has also been studied by immobilization on several substrates such as Ca-alginate beads, phosphocellulose, and triethylaminoethyl cellulose [77, 85]. In every method of pullulan production, it is necessary to maintain several culture conditions such as carbon source, nitrogen source, pH, temperature, and other factors such as minerals, vitamins, aeration, agitation, etc. These factors that affect pullulan production are discussed as follows.

4.3.1 Carbon Source

Carbon plays a major role since it is an essential structural component of framework for fungal cell. All fungal species utilize carbon for their growth, and their utilization depends on the carbon compound. While some fungi use simpler sugars, other species utilize special sugars [7, 29]. Various studies have been performed on the effect of sugars on pullulan production since the strains producing pullulan are easily affected by excess sugar present in the medium; therefore, new strains having the ability to withstand too much sugar have to be identified. While several researches have been carried out in identifying new strains for pullulan production, other works on finding apt carbon source have also been performed. The different carbon sources studied for better production of pullulan are listed in Table 4.2.

Table 4.2 Carbon sources used for pullulan production

S. No.	Carbon source	Reference
1.	Glucose	14, 21, 34, 63, 84
2.	Maltose	10
3.	Xylose	21
4.	Sucrose	13, 14, 21, 61, 84
5.	Dextrose	34
6.	Hemicellulose	6
7.	Starch	91
8.	Jaggery	20
9.	Potato starch hydrolysate	1

(Continued)

Table 4.2 *(Continued)*

S. No.	Carbon source	Reference
10.	Grape skin pulp	32
11.	Molasses	33
12.	Carob pod	55
13.	De-oiled rice bran extract	71
14.	Fructose	10, 14, 21, 91
15.	Cheese whey	41
16.	Peat hydrolysate	8
17.	Olive oil	92
18.	Lactose	14, 21, 63
19.	Inulin	64
20.	Spent grain liquor	58
21.	Beet molasses	57
22.	Corn meal hydrolysate	31
23.	Spent sulfite liquor	95

4.3.2 Nitrogen Source

Nitrogen is an important source for protein production and other vital source in fungal cells. The carbon-to-nitrogen ratio has been found to influence the rate of assimilation of nitrogen into microbial biomass [7, 47, 96]. Change in the nitrogen state of the medium also alters pH, which is a crucial factor in cell morphology and pullulan synthesis [39, 44]. Various nitrogen sources have been studied for pullulan production.

Ammonium sulfate is the predominantly used nitrogen source for pullulan production. But in the later years, several nitrogen sources were identified, such as yeast extract [21, 66], corn steep liquor [52, 81], soytone [52, 81], soybean pomace [62], phenylalanine [2], histidine [2, 21], glycine [2], arginine [2], alanine [2], monosodium glutamate [2], $NaNO_2$ [2, 21], $NaNO_3$ [2, 21, 51, 66], KNO_3 [2], NH_4NO_3 [2], $(NH_4)_2SO_4$ [2, 21, 51, 66, 81], ammonium tartarate [2], $(NH_4)_2CO_3$, tryptone [52], peptone [51, 52, 90], casamino acids [52], soybean meal hydrolysate [91], etc. The highest pullulan production was found with peptone at the rate of 72.56 g/L [90], then with yeast extract at the rate of 65 g/L, and finally with soybean meal hydrolysate at the rate of 59.8 g/L [91]. Apart from these nitrogen sources, $NaNO_2$ was found to increase the synthesis of UDP–glucose pyrophosphorylase, an enzyme that plays a vital role in the synthesis of pullulan polysaccharide [35]. The usage of casamino acids has led to the morphological alterations of strain, thereby affecting pullulan production. Therefore, except casamino acids, their nitrogen sources were found to have better pullulan production.

4.3.3 pH

pH plays a vital role in pullulan production. Heald and Kristiansen [26] proposed that the rate of pullulan synthesis is independent of pH. But several works on the effect of pH on pullulan production showed that pH is important for the growth of the pullulan-producing strain, thereby affecting pullulan production [26]. It has also been identified that pH alters the morphology of pullulan-producing strains. The effect of pH on pullulan production with sucrose as carbon source was studied by Lacroix et al. [39]. He found that at pH 2, polysaccharide production was very low at a concentration of 1 g/L, but it increased to pH 14–15 g/L at pH 5.5 [39]. They also performed an experiment maintaining bistaged pH in which an initial pH of 2 was maintained for the first 4 days, which is called the first stage and the pH was increased to 4–5 on the fourth day, which was maintained for 7 days. This 7-day period was called the second stage. When this kind of pH adjustment was maintained, the strains were found to produce more pullulan at a concentration of 26–27 g/L, which was almost an increase of 44–48% pullulan production.

This variable production of pullulan at different pH was attributed to the morphology of the pullulan-producing strain. At the initial pH of 2–2.5, pullulan was found to exist in M phase where it has yeast-like appearance. At this lower initial pH, the morphology of the strains was maintained throughout their growth and pullulan production. But at the initial pH of 6, strains transformed their morphology from yeast-like appearance to filamentous appearance. This change to filamentous appearance was called Y phase. So it has been identified that pullulan production is at its best when the organism is in the Y phase where the elaboration of pullulan is efficient. Therefore, pH 6–6.5 has been considered optimum for pullulan production [10, 50]. Apart from the effect of pH on the morphology of pullulan-producing strains, it has its effects also on the molecular weight of pullulan. At pH less than 5, low-molecular-weight pullulan was synthesized, which was attributed toward the synthesis of pullulan-degrading enzymes at this pH. Whereas at a higher initial pH of 6–6.5, a higher-molecular-weight pullulan was synthesized [42].

Later studies on the effect of pH on pullulan production were carried out with various carbon and nitrogen sources. When using sucrose as the carbon source and yeast extract as the nitrogen source, the concentration of pullulan was found to increase at pH 6.5, which was obtained after 5 days of incubation with initial pH 2. But pullulan production was found to decrease at pH 7–7.5. Pullulan production at low level was also detected in the absence of yeast extract. Similarly, when using corn syrup as the carbon source and yeast extract as the nitrogen source, at low pH of 2, pullulan production was very low both in presence and absence of yeast extract. But at the initial pH of 6.5–7, pullulan production increased and it was optimal at pH 5–7. The production of pullulan was found to decrease abruptly at initial pH 7.5 [52, 86].

Lee and Yoo (1993) obtained better pullulan production at an optimal pH of 6.0 [42]. Wang et al. (2013) found that although maximum biomass of 14.45 g/L was produced at pH 6.5, the maximum pullulan concentration of 26.8 g/L was obtained only at pH of 3.8 [80]. Even in the commercial production of pullulan by Tsujisaka and Mitsuhashi [76], pH was maintained at 6.5. When using coconut water and coconut milk as the carbon source, the initial pH 7 was found to be optimum in pullulan production [74, 76]. In another study by Singh et al. [67], sugar cane juice was used as the carbon source and with this source, initial pH 5 was found to enhance pullulan production profoundly. Though pullulan production was found to be affected by pH, its main property, viscosity, was unaffected by pH values [27, 67].

4.3.4 Temperature

The optimal temperature for pullulan production was found to be around 24–35°C [30, 43, 75, 94]. Similar to pH, temperature also has two-way roles in pullulan production and growth of pullulan-producing strain since both the temperatures vary greatly. In a work by West and Reed-Hamer [86], optimal temperature for the production of pullulan was found to be 26°C, whereas the optimal temperature for the growth of *A. pullulans* was found to be 33°C [87]. Therefore, a two-staged temperature has to be maintained for growing *A. pullulans* and to produce pullulan from them [52]. Similar works were performed by Wu et al. [88] and Xia et al. [89] in which the initial concentration of 32°C was maintained for pullulan growth, which was then reduced to 26°C for pullulan production.

4.3.5 Other Factors

Other fermentation factors such as agitation, aeration, and incubation time are also considered for pullulan production. Incubation time was studied by several researchers to find maximum production. A wide variety of incubation times, such as 24 h, 46 h, 60 h, 72 h, 144 h, 168 h, were studied with varying different sources and the optimum incubation time varied with varying sources and strains [16, 18, 23]. In addition to this, aeration and agitation were also studied for pullulan production. An impeller speed, i.e., agitation of 200 rpm, was usually followed, but studies by Gniewosz et al. [23] and Dixit et al. [18] on varying agitation speeds, such as 200 rpm, 400 rpm, 600 rpm, 800 rpm, and 1000 rpm, showed optimum agitation at 800 rpm and 600 rpm, respectively. In previous studies, aeration was not depicted as a factor that affects pullulan production [22, 24, 56]. But the study on aeration by Dixit et al. [18] showed that high aeration improved mass transfer during fermentative production of pullulan.

4.4 Conclusion

Since pullulan is an efficient polysaccharide known for its several applications in pharmaceutical and food industry, its commercial production must be carried out in an efficient way for better production at low cost. Several factors affecting pullulan production have been discussed, and it is now proposed that various sources and strains require different optimal conditions for pullulan production. Therefore, it is necessary to make standard optimal fermentation conditions for pullulan production that involve low cost of sources and high production of pullulan.

References

1. An, C., Ma, S., Chang, F., and Xue, W. (2017). Efficient production of pullulan by *Aureobasidium pullulans* grown on mixtures of potato starch hydrolysate and sucrose. *Brazilian J. Microbiol.*, **48**(1), 180–185.
2. Auer, D. P. F. and Seviour, R. J. (1990). Influence of varying nitrogen sources on polysaccharide production by *Aureobasidium pullulans* in batch culture. *Appl. Microbiol. Biotechnol.*, **32**, 637–644.
3. Badr-Eldin, S. M., El-Tayeb, O. M., El-Masry, H. G., Mohamad, F. H. A., and Abd El-Rahman, O. A. (1994). Polysaccharide production by *Aureobasidium pullulans*: Factors affecting polysaccharide formation. *World J. Microbiol. Biotechnol.*, **10**, 423–426.
4. Behrens, U. and Lohse, R. (1977). Investigations for the production of pullulan. *Die Nahrung*, **21**, 199–204.
5. Bender, H., Lehmann, J., and Wallenfels, K. (1959). Pullulan, in extracellulares Glucan von Pullularia pullulans. *Biochim. Biophys. Acta*, **36**, 309–316.
6. Biely, P., Kratky, Z., Petrakova, E., and Bauer, S. (1978). Growth of *Aureobasidium pullulans* on waste water hemicelluloses. *Folia Microbiologica*, **24**, 328–333.
7. Bilgrami, K. S. and Verma, R. N. (1978). *Physiology of Fungi*. Vikas Publishing House Pvt. Ltd., New Delhi, 493.
8. Boa, J. M. and Leduy, A. (1984). Peat hydrolysate medium optimization for pullulan production. *Appl. Environ. Microbiol.*, **48**(1), 26–30.
9. Bouveng, H. O., Kiessling, H., Lindberg, B., and McKay, J. (1962). Polysaccharides elaborated by *Pullularia pullulans*. 1. The neutral glucan synthesised from sucrose solutions. *Acta Chern. Scund.*, **16**, 615–622.
10. Catley, B. J. (1971). Utilization of carbon sources by *Pullularia pullulans* for the elaboration of extracellular polysaccharides. *Appl. Microbiol.*, **22**(4), 641–649.
11. Chatap, V. K., Maurya, A. R., Deshmukh, P. K., and Zawar, L. R. (2013). Formulation and evaluation of nisoldipne sublingual tablets using pullulan and chitosan for rapid oromucosal absorption. *Adv. Pharmacol. Pharm.*, **1**, 18–25.
12. Chen, G., Wang, J., Su, Y., Zhu, Y., Zhang, G., Zhao, H., Liu, H., Yang, Y., Nian, R., Zhang, H., Wei, Y., and Xian, M. (2017). Pullulan production from synthetic medium by a new mutant of *Aureobasidium pullulans*. *Prep.*

Biochem. Biotechnol., **47**(10), 963–969, DOI:10.1080/10826068.2017.1350979.

13. Chen, J., Wu, S., and Pan, S. (2012). Optimization of medium for pullulan production using a novel strain of *Aureobasidium pullulans* isolated from sea mud through response surface methodology. *Carbohydr. Polym.*, **87**(1), 771–774.
14. Cheng, K. C., Demirci, A., and Catchmark, J. M. (2011). Evaluation of medium composition and fermentation parameters on pullulan production by *Aureobasidium pullulans*. *Food Sci. Technol. Int.*, **17**(2), 99–109, DOI:10.1177/1082013210368719.
15. Chi, Z. and Zhao, S. (2003). Optimization of medium and cultivation conditions for pullulan production by a new pullulan-producing yeast strain. *Enzyme Microb. Technol.*, **33**, 206–211.
16. Choudhury, A. R., Bhattacharjee, P., and Prasad, G. S. (2013). Development of suitable solvent system for downstream processing of biopolymer pullulan using response surface methodology. *Plos One*, **8**(10), e77071.
17. Delben, F., Forabosco, A., Guerrini, M., Liut, G., and Torri, G. (2006). Pullulans produced by strains of *Cryphonectria parasitica*-II. Nuclear magnetic resonance evidence. *Carbohydr. Polym.*, **63**, 545–554.
18. Dixit, P., Mehta, A., Gahlawat, G., Prasad, G. S., and Choudhury, A. R. (2015). Understanding the effect of interaction among aeration, agitation and impeller positions on mass transfer during pullulan fermentation by *Aureobasidium pullulans*. *RSC Adv.*, **5**, 38984–38994.
19. Fraser, C. G. and Jennings, H. J. (1971). A glucan from *Tremella mesenterica* NRRL-Y6158. *Can. J. Chem.*, **49**, 1804–1807.
20. Ganduri, V. S. R. K., Sambasiva Rao, K. R. S., Mangamuri, U. K., Vijaya Lakahmi, M., and Poda, S. (2016). Production of pullulan using jaggery as substrate by *Aureobasidium pullulans* MTCC 2195. *Curr. Trends Biotechnol. Pharm.*, **10**(2), 153–160.
21. Gaur, R. and Singh, R. (2010). Optimization of physico-chemical and nutritional parameters for pullulan production by a mutant of thermotolerant *Aureobasidium pullulans* in fed batch fermentation process. *Afr. J. Biotechnol.*, **9**(43), 7322–7330.
22. Gibbs, P. and Seviour, R. (1996). Does the agitation rate and/or oxygen saturation influence exopolysaccharide production by *Aureobasidium pullulans* in batch culture? *Appl. Microbiol. Biotechnol.*, **46**, 503–510.

23. Gniewosz, M., Kraśniewska, K., and Synowiec, A. (2013). The effect of agitation on pullulan production by a white mutant *Aureobasidium pullulans* B-1 in batch culture. *EJPAU*, **16**(2), 1–11.

24. Goksungur, Y., Dagbaglı, S., Uçan, A., and Guvenç, U. (2005). Optimization of pullulan production from synthetic medium by *Aureobasidium pullulans* in a stirred tank reactor by response surface methodology. *J. Chem. Technol. Biotechnol.*, **80**, 819–827.

25. Goksungur, Y., Uzunoullar, P., and Dabal, S. (2011). Optimization of pullulan production from hydrolysed potato starch waste by response surface methodology. *Carbohydr. Polym.*, **83**(3), 1330–1337.

26. Heald, P. J. and Kristiansen, B. (1985). Synthesis of polysaccharide by yeast-like forms of *Aureobasidium pullulans. Biotechnol. Bioengineer.*, **27**, 1516–1519.

27. Hijiya, H. and Shiosaka, M. (1975). *Adhesives and Pastes*. US Patent Office, Pat. No. 3 873 333.

28. https://www.chemicalbook.com/ChemicalProductProperty_US_CB8141488.aspx

29. Husian, S. S. and Zamir, K. (1970). Studies on nutrition of Fungi. *Pak J. Bot.*, **3**(2), 79–82.

30. Imshenetskii, A. A., Kondrat'eva, T. F., and Smut'ko, A. N. (1981). Influence of the acidity of the medium, conditions of aeration, and temperature on pullulan biosynthesis by polyploid strains of *Pullularia (Aureobasidium) pullulans. Mikrobiologiia*, **50**, 471–475.

31. Imshenetskii, A. A., Kondrat'eva, T. F., Dvadtsamova, E. A., and Vorontsova, N. N. (1985). Activity of pullulan synthesis by diploid culture of *Pullularia pullulans* on media with different carbon sources. *Mikrobiologiia*, **54**, 927–929.

32. Israilides, C., Scanlon, B., Smith, A., Jumel, K., and Harding, S. E. (1994). Characterization of pullulans produced from agro-industrial wastes. *Carbohydr. Polym.*, **25**, 203–309.

33. Israilides, C., Smith, A., and Bambalov, G. (1993). Production of pullulan from agroindustrial wastes. *Proceedings of the Sixth European Congress on Biotechnology* **2**, 362 Firenze, Italy.

34. Jadhav, D. M. and Gawai, D. U. (2012). Effect of different nutrient sources on biomass production of phylloplane yeast *Aureobasidium pullulans* (De Bary). *Int. Res. J. Biological Sci.*, **1**(8), 85–87.

35. Jiang, L., Wu, S., and Kim, J. M. (2011). Effect of different nitrogen sources on activities of UDPG-pyrophosphorylase involved in pullulan

synthesis and pullulan production by *Aureobasidium pullulans*. *Carbohydr. Polym.*, **86**(2), 1085–1088.

36. Kageyama, S., Wada, H., Muro, K., Niwa, Y., Ueda, S., Miyata, H., Takiguchi, S., Sugino, S. H., Miyahara, Y., Ikeda, H., Imai, N., Sato, E., Yamada, T., Osako, M., Ohnishi, M., Harada, N., Hishida, T., Doki, Y., and Shiku, H. (2013). Dose-dependent effects of NY-ESO-1 protein vaccine complexed with cholesteryl pullulan (CHP-NY-ESO-1) on immune responses and survival benefits of esophageal cancer patients. *J. Transl. Med.*, **11**, 246, 1–10.

37. Kong, I. G., Sato, A., Yuki, Y., Nochi, T., Takahashi, H., Sawada, S., Mejima, M., Kurokawa, S., Okada, K., Sato, S., Briles, D. E., Kunisawa, J., Inoue, Y., Yamamoto, M., Akiyoshi, K., and Kiyono, H. (2013). Nanogel-based PspA intranasal vaccine prevents invasive disease and nasal colonization by *Streptococcus pneumonia*. *Infect. Immun.*, **81**, 1625–1634.

38. Kong, S. H., Noh, Y. W., Suh, Y. S., Park, H. S., Lee, H. J., Kang, K. W., Kim, H. C., Lim, Y. T., and Yang, H.-K. (2015). Evaluation of the novel near-infrared fluorescence tracers pullulan polymer nanogel and indocyanine green/γ-glutamic acid complex for sentinel lymph node navigation surgery in large animal models. *Gastric Cancer*, **18**, 55–64.

39. Lacroix, C., LeDuy, A., Noel, G., and Choplin, N. (1985). Effect of pH on the batch fermentation of pullulan from sucrose medium. *Biotechnol. Bioeng.*, **27**, 202–207.

40. Leathers, T. D. (2003). Biotechnological production and applications of pullulan. *Appl. Microbiol. Biotechnol.*, **62**, 468–473.

41. LeDuy, A., Yarmoff, J.-J., and Chagraoui, A. (1983). Enhanced production of pullulan from lactose by adaptation and by mixed culture techniques. *Biotechnol. Lett.*, **5**, 49–54.

42. Lee, K. Y. and Yoo, Y. J. (1993). Optimization of pH for high molecular weight pullulan. *Biotechnol. Lett.*, **15**(10), 1021–1024.

43. McNeil, B. and Kristiansen, B. (1990). Temperature effects on polysaccharide formation by *Aureobasidium pullulans* in stirred tanks. *Enzyme Microb. Technol.*, **12**, 521–526.

44. McNeil, B., Kristiansen, B., and Seviour, R. J. (1989). Polysaccharide production and morphology of *Aureobasidium pullulans* in continuous culture. *Biotechnol. Bioeng.*, **33**(9), 1210–1212.

45. Mishra, M. (2017). Major problems addressed in pullulan production: A review. *Adv. Biotech. Micro.*, **6**(5): AIBM.MS.ID.555696.

46. Moscovici, M., Ionescu, C., Oniscu, C., Fotea, O., Protopopescu, P., Hanganu, L. D., and Vitan, C. (1996). Improved exopolysaccharide

production in fed-batch fermentation of *Aureobasidium pullulans*, with increased impeller speed. *Biotechnol. Lett.*, **18**, 787–790.

47. Narasimha, G., Sridevi, A., Buddolla, V., Subhosh, C. M., and Rajaseekhar, R. B. (2006). Nutrient effect on production of cellulolytic enzymes by *Aspergillus niger. Afr. J. Biotechnol.*, **5**(5), 472–476.

48. Oğuzhan, P. and Yangılar, F. (2013). Pullulan: Production and usage in food industry. *Afr. J. Food Sci. Technol.*, **4**(3), 57–63.

49. Oliva, E. M., Cirelli, A. F., and De Lederkremer, R. M. (1986). Characterization of a pullulan in *Cyttaria darwinii*. *Carbohydr. Res.*, **158**, 262–267.

50. Ono, K., Yasuda, N., and Ueda, S. (1977). Effect of pH on pullulan elaboration by *Aureobasidium pullulans* S-lt. *Agric. Biol. Chem.*, **41**(11), 2113–2118.

51. Prasongsuk, S., Berhow, M. A., Dunlap, C. A., Weisleder, D., Leathers, T. D., Eveleigh, D. E., and Punnapayak, H. (2007). Pullulan production by tropical isolates of *Aureobasidium pullulans*. *J. Ind. Microbiol. Biotechnol.*, **34**, 55–61, DOI:10.1007/s10295-006-0163-7.

52. Reed-Hamer, B. and West, T. P. (1994). Effects of complex nitrogen sources on pullulan production relative to carbon source. *Microbios*, **80**, 83–90.

53. Reeslev, M. and Jensen, B. (1995). Influence of Zn^{2+} and Fe^{3+} on polysaccharide production and mycelium/yeast dimorphism of *Aureobasidium pullulans* in batch cultivations. *Appl. Microbiol. Biotechnol.*, **42**, 910–915.

54. Reis, R. A., Tischer, C. A., Gorrin, P. A. J., and Iacomini, M. (2002). A new pullulan and a branched (1→3)-, (1→6)-linked b-glucan from the lichenised ascomycete *Teloschistes flavicans*. *FEMS Microbiol. Lett.*, **210**, 1–5.

55. Rollkas, T. and Biliaderis, C. (1995). Evaluation of carob pod as a substrate for pullulan production by *Aureobasidium pullulans*. *Appl. Biochem. Biotechnol.*, **55**, 27–44.

56. Roukas, T. and Mantzouridou, F. (2001). Effect of the aeration rate on pullulan production and fermentation broth rheological properties in an airlift reactor. *J. Chem. Technol. Biotechnol.*, **76**, 371–376.

57. Roukas, T. and Liakopoulou-Kriakides, M. (1999). Production of pullulan from beet molasses by *Aureobasidium pullulans* in a stirred tank fermentor. *J. Food Eng.*, **40**, 89–94.

58. Roukas, T. C. (1999). Pullulan production from brewery wastes by *Aureobasidium pullulans*. *World J. Microbiol. Biotechnol.*, **15**, 447–450.

59. Sawada, S. I., Sasaki, Y., Nomura, Y., and Akiyoshi, K. (2011). Cyclodextrin-responsive nanogel as an artificial chaperone for horseradish peroxidase. *Colloids Polym. Sci.*, **289**, 685–691.

60. Schuster, R., Wenzig, E., and Mersmann, A. (1993). Production of the fungal exopolysaccharide pullulan by batch-wise and continuous fermentation. *Appl. Microbiol. Biotechnol.*, **39**, 155–158.

61. Sena, R. F., Costelli, M. C., Gibson, L. H., and Coughlin, R. W. (2006). Enhanced production of pullulan by two strains of *A. pullulans* with different concentrations of soybean oil in sucrose solution in batch fermentations. *Braz. J. Chem. Eng.*, **23**(4), 507–515.

62. Seo, H. P., Son, C. W., Chung, C. H., Jung, D., Kim, S. K., Gross, R. A., Kaplan, D. L., and Lee, J.-W. (2004). Production of high molecular weight pullulan by *Aureobasidium pullulans* HP-2001 with soybean pomace as a nitrogen source. *Bioresour. Technol.*, **95**, 292–299.

63. Sheoran, S. K., Dubey, K. K., Tiwari, D. P., and Singh, B. P. (2012). Directive production of pullulan by altering cheap source of carbons and nitrogen at 5 L bioreactor level. *ISRN Chem. Eng.*, 2012, Article ID 867198, 1–5, DOI:10.5402/2012/867198.

64. Shin, Y. C., Kim, Y. H., Lee, H. S., Cho, S. J., and Byun, S. M. (1989). Production of exopolysaccharide pullulan from inulin by mixed culture of *Aureobasidium pullulans* and *Kluyveromyces fragilis*. *Biotechnol. Bioeng.*, **33**, 129–133.

65. Shingel, K. I. (2004). Current knowledge on biosynthesis, biological activity, and chemical modification of the exopolysaccharide pullulan. *Carbohydr. Res.*, **339**, 447–460.

66. Singh, R., Gaur, R., Bansal, S., Jamal, F., Pandey, P. K., Tiwari, S., Sarsaiya, S., Mishra, S., Chaturvedi, N., Singh, D. P., Gaur, M. K., and Vazhan Arasu, F. G. (2017). Production of pullulan from a high yielding strain of *Aureobasidium pullulans* in non-stirred flask-type fermentation system. *J. Microbiol. Biotech. Res.*, **7**(1): 26–32, DOI: 10.24896/ jmbr.2017715.

67. Singh, R., Gaur, R., Pandey, P., Jamal, F., Pandey, L., Singh, S., Kewat, H., Tiwari, S., Biswas, P., and Gaur, M. (2018). A novel media optimized for production of pullulan in flask type fermentation system. *Int. J. Curr. Microbiol. Appl. Sci.*, **7**, 53–61. 10.20546/ijcmas.2018.704.007.

68. Singh, R., Gaur, R., Tiwari, S., and Gaur, M. K. (2012). Production of pullulan by a thermo tolerant *Aureobasidium pullulans* strain in non-stirred fed batch fermentation process. *Braz. J. Microbiol.*, **43**(3), 1042–1050.

69. Singh, R. S. and Kaur, N. (2015). Microbial biopolymers for edible film and coating applications. In: Nawani, N. N., Khetmalas, M., Razdan, P.

N., Pandey, A. (Eds.), *Advances in Biotechnology*. India: IK International Publishing, pp. 187–216.

70. Singh, R. S. and Kaur, N. (2018). Biochemical and molecular characterization of a new pullulan producer *Rhodosporidium paludigenum* PUPY-06. *J. Appl. Biol. Biotechnol.*, **6**(1), 28–37.
71. Singh, R. S. and Kaur, N. (2019). Understanding response surface optimization of medium composition for pullulan production from de-oiled rice bran by *Aureobasidium pullulans. Food Sci. Biotechnol.*, **28**(5), 1507–1520, https://doi.org/10.1007/s10068-019-00585-w.
72. Singh, R. S., Kaur, N., and Kennedy, J. F. (2015). Pullulan and pullulan derivatives as promising biomolecules for drug and gene targeting. *Carbohydr. Polym.*, **123**, 190–207.
73. Singh, R. S., Saini, G. K., and Kennedy, J. F. (2008). Pullulan: Microbial sources, production and applications. *Carbohydr. Polym.*, **73**(4), 515–531.
74. Thirumavalavan, K., Manikkadan, T. R., and Dhanasekar, R. (2009). Pullulan production from coconut by-products by *Aureobasidium pullulans. Afr. J. Biotechnol.*, **8**(2), 254–258.
75. Torzilli, A. P., Vinroot, S., and West, C. (1985). Interactive effect of temperature and salinity on growth and activity of a salt marsh isolate of *Aureobasidium pullulans. Mycologia*, **77**(2), 278–284.
76. Tsujisaka, Y. and Mitsuhashi, M. (1993). Pullulan. In: Whistler, R. L. and BeMiller, I. N. (Eds.), *Industrial Gums. Polysaccharides and Their Derivatives*, 3rd edn. San Diego: Academic Press, pp. 447–460.
77. Ürküt, Z., Dağbağli, S., and Göksungur, Y. (2007). Optimization of pullulan production using Ca-alginate-immobilized *Aureobasidium pullulans* by response surface methodology. *J. Chem. Technol. Biotechnol.*, **82**, 837–846.
78. Vijayendra, S. V. N., Bansal, D., Prasad, M. S., and Nand, K. (2001). Jaggery: A novel substrate for pullulan production by *Aureobasidium pullulans* CFR-77. *Proc. Biochem.*, **37**, 359–364.
79. Waksman, N., De Lederkremer, R. M., and Cerezo, A. S. (1977). The structure of an a-D-glucan from *Cyttaria harioti* Fischer. *Carbohydr. Res.*, **59**, 505–515.
80. Wang, D., Yu, X. and Gongyuan, W. (2013). Pullulan production and physiological characteristics of *Aureobasidium pullulans* under acid stress. *Appl. Microbiol. Biotechnol.*, **97**(18), 8069–8077, doi: 10.1007/s00253-013-5094-4.

81. West, T. P. and Strohfus, B. (1999). Effect of nitrogen source on pullulan production by *Aureobasidium pullulans* grown in a batch bioreactor. *Microbios*, **99**, 147–159.

82. West, T. P. and Reed-Hamer, B. (1991). Ability of *Aureobasidium pullulans* to synthesize pullulan upon selected sources of carbon and nitrogen. *Microbios*, **67**, 117–124.

83. West, T. P. and Reed-Hamer, B. (1992). Influence of vitamins and mineral salts upon pullulan synthesis of *Aureobasidium pullulans*. *Microbios*, **71**, 115–123.

84. West, T. P. (2011). Effect of carbon source on polysaccharide production by alginate-entrapped *Aureobasidium pullulans* ATCC 42023 cells. *J. Basic Microbiol.*, **51**, 673–677.

85. West, T. P. (2012). Pullulan production by *Aureobasidium pullulans* ATCC 201253 cells adsorbed onto cellulose anion and cation exchangers. *ISRN Microbiol.*, 2012, Article ID 140951, 1–4.

86. West, T. P. and Reed-Hamer, B. (1993). Effect of pH on pullulan production relative to carbon source and yeast extract composition of growth medium. *Microbios*, **75**, 75–82.

87. West, T. P. and Reed-Hamer, B. (1993). Effect of temperature on pullulan production in relation to carbon source. *Microbios*, **75**, 261–268.

88. Wu, S., Chen, H., Jin, Z., and Tong, Q. (2010). Effect of two-stage temperature on pullulan production by *Aureobasidium pullulans*. *World J. Microbiol. Biotechnol.*, **26**, 737–741, DOI:10.1007/s11274-009-0231-z.

89. Xia, Z., Wu, S., and Pan, S. (2011). Effect of two-stage controlled pH and temperature on pullulan production by *Aureobasidium pullulans*. *Carbohydr. Polym.*, **86**(4), 1814–1816.

90. Xie, Y., Ma, X., Gao, Y., and Lu, X. (2013). Optimization of fermentation medium for pullulan production by *Aureobasidium pullulans* A225 using Plackett–Burman and response surface methodology. *Adv. Mater. Res. Online*, **785–786**, 279–286, DOI:10.4028/www.scientific.net/AMR.785-786.279.

91. Yang, J., Zhang, Y., Zhao, S., Zhou, Q., Xin, X., and Chen, L. (2018). Statistical optimization of medium for pullulan production by *Aureobasidium pullulans* NCPS2016 using fructose and soybean meal hydrolysates. *Molecules*, **23**(6), E1334. 1–16, DOI:10.3390/molecules23061334.

92. Youssef, F., Biliaderis, C. G., and Roukas, T. (1998). Enhancement of pullulan production by *Aureobasidium pullulans* in batch culture using

olive oil and sucrose as carbon sources. *Appl. Biochem. Biotechnol.*, **74**, 13–30.

93. Yuen, S. (1974). Pullulan and its applications. *Process Biochem.*, **9**(9), 7–9, 22.
94. Zajic, J. E. (1967). Process for preparing a polysaccharide flocculating agent. US Patent Office, Pat. No. 3 320 136.
95. Zajic, J. E., Ho, K. K., and Kosaric, N. (1979). Growth and pullulan production by *Aureobasidium pullulans* on spent sulphite liquor. In: Underkofler, L. A. (Ed.), *Developments in Industrial Microbiology*. Arlington, VA: Society for Industrial Microbiology, pp. 631–639.
96. Zibilske, L. M. (1999). Composting of organic wastes. In: Sylvia, D. M., Fuhrmann, J. J., Hartl, P. G., and Zuberer, D. A. (Eds.), *Principles and Applications of Soil Microbiology*. New Jersey, USA: Prentice Hall, 484.

Chapter 5

Pullulan: Properties and Applications

Showkat Ali Ganie,[a,b] Tariq Ahmad Mir,[b]
Akbar Ali,[b] and Qing Li[a]
[a]*State Key Laboratory of Silkworm Genome Biology,*
Chongqing Engineering Research, Center for Biomaterial Fiber
and Modern Textile, College of Textile and Garments,
Southwest University, Chongqing 400715, PR China
[b]*Department of Chemistry, Jamia Millia Islamia,*
New Delhi 110025, India
alishowkat443@gmail.com

Pullulan is an extracellular polysaccharide polymer produced by the fungus *Aureobasidium pullulans*. Structurally, it is composed of maltotriose units, and each maltotriose unit contains three glucose units linked through α-1,4 glycosidic linkages. The consecutive maltotriose units in the structure are connected through α-1,6 glycosidic bonds with each other. Currently, pullulan is extensively used in the food industry as a low-calorie food additive; it also possesses good moisture retention capability and oxygen barrier property. In this chapter, we have highlighted the structure, properties, and applications of pullulan in different fields, such as cosmetic industry, drug delivery, environmental applications, tissue engineering, biomedical applications, etc. Finally, the various

Pullulan: Processing, Properties, and Applications
Edited by Shakeel Ahmed and Aisverya Soundararajan
Copyright © 2021 Jenny Stanford Publishing Pte. Ltd.
ISBN 978-981-4877-17-6 (Hardcover), 978-1-003-05694-2 (eBook)
www.jennystanford.com

challenges that occurred in the application of pullulan and future perspectives of pullulan at the industrial scale in the development of new technological aspects are also discussed.

5.1 Introduction

The comprehensive assortment of exopolysaccharides produced from microorganisms has distinctive physical and chemical characteristics [1]. These exopolysaccharides can be classified as homopolysaccharides, which are commonly neutral glucans and heteropolysaccharides, which are polyanionic because of the presence of uronic acid [2–5]. The unique and unrivaled physical properties make the microbial exopolysaccharides auspicious substitutes for plant polysaccharides. Depending on their morphological confinement, the polysaccharides produced microbially can be intracellular and extracellular [6, 7]. The plant-producing natural polysaccharides are replaced by microbial polymers due to their industrial importance. Some of the biopolymers produced microbially are haloferax exopolysaccharides, pullulan, levan, kefiran, gellan, xanthan, polyhydroxyalkanoate, bacterial cellulose, etc. [8]. Pullulan is one of such significant polymers delivered by the yeast-like fungus *A. pullulans*. Microbial exopolysaccharides broadly contemplated these days incorporate pullulan from *A. pullulans*, succinoglycan from *Rhizobium*, xanthan from *Xanthomonas campesteris*, bacterial alginates from *Pseudomonas* spp. and *Azotobacter vinelandii*, with potential mechanical applications [9]. These macromolecular groups of polysaccharides have various applications in energy production, food, chemical, and pharmaceutical industries. A couple of fungal exopolysaccharides with fascinating mechanical properties are notable. Pullulan is one of such financially rising biopolymers and incorporated by a yeast-like fungus *A. pullulans*. It is a water-dissolvable irregular loop glucan that fills in as a worldview for the conduct of aqueous polysaccharides [10–12]. As of now, just a few exocellular fungal alpha glucans are accounted for in the literature. Pullulan is the best known alpha-linked glucan formed by the polymorphic organism *A. pullulans*. In 1938, Bauer was a pioneer who noticed the polysaccharide production by *A. pullulans* [13]. First isolated and distinguished by Bernier in 1958 from the culture broths of *A. pullulans*, pullulan has become the subject of a growing study attempt [14]. Bender, Lehmann, and Wallenfels, in

1959, performed a detailed analysis of this novel polysaccharide and named it "pullulan" [15]. Various research gatherings that worked for clarifying the pullulan structure announced that the structure of pullulan could not be described as comprising exclusively of α-(1-4) and α-(1-6) linkages, in light of the fact that a little extent of α-(1-3) linkages are additionally detailed in the structure [16, 17].

In the early 1960s, Wallenfels et al. settled the pullulan structure to prove that pullulan could be extended and have groups of maltosyl or glucosyl rather than maltotriosyl residues [18–21]. Wallenfels et al. in 1961 reported the action of enzyme pullulanase on pullulan [19]. The enzyme explicitly hydrolyzes the α-(1-6) connections in pullulan yielding quantitatively the oligomers (G3)n, where "G3" speaks to α -(1-4)-connected repeating trimer maltotriose, and "n" is the quantity of trimeric repeating units in the oligomer. Catley et al., in 1986, assumed that there might be no notable pullulan structure as announced by Catley in 1970 [22]. Some of the interesting reviews on the structural elucidation of pullulan that have been done so far include those by Gibbs [23, 24].

Since 1976, the Hayashibara Company Limited in Okayama, Japan, has been manufacturing pullulan. Hayashibara balanced the growth conditions of the source fungus to create pullulan of specific molecular weight and determination. These incorporate food-grade (assigned as PF) and deionized (PI) items with a mean molecular weight of 100,000 (PI-10 and PF-10) or 200,000 (PI-20 and PF-20). Hayashibara marketed the pullulan films in 1982. Today, this organization is the key commercial maker of pullulan. Hayashibara Biochemical Labs (HBL) has conceded Pfizer an overall permit to popularize and market film-based oral consideration items containing HBL's licensed pullulan and has initiated providing pullulan powder to the Warner-Lambert Company, an auxiliary of Pfizer Inc. Pfizer has presented the film structure oral care item in Canada under the brand name Listerine.

5.2 Structure of Pullulan

Polysaccharides delivered by microorganisms have a property that is exceptionally helpful in different mechanical applications [25]. Another fungal exopolysaccharide with interesting mechanical characteristics has been well established. Pullulan is an extracellular and unbiased microbial polysaccharide delivered by *A. pullulans* in starch and sugar cultures [26–28]. Pullulan is a linear, neutral [29], nonionic [30], and water-soluble polysaccharide [31] made out of α-(1, 6) repeated maltotriose units by means of α-(1, 4) glycosidic bond (Fig. 5,1) [32]. The chemical formula of pullulan is $C_6H_{10}0_5$, and pullulan has an optical rotation of 192° in 1 g/dl solution [24]. Its linear structure includes subunits of maltotetraose. Its structure is like starch [9]. The average molecular weight of pullulan ranges from 5×10^3 to 2×10^6 Da and depends on the circumstances of strain and culture [32]. The structure of pullulan was settled in the 1960s [33]. It has been demonstrated that enzyme pullulanase specifically hydrolyzes α-(1-6) linkages in pullulan, yielding quantitatively the oligomer α-(1-4)-linked repeating trimer maltotriose. The maximum extent of subunits may be 7% and can be distributed all through the molecule [34]. Sometimes pullulan partial acid hydrolysis also produces maltose, isomaltose, and panose disopanose [15, 17]. The branched structure of pullulan has maltosylorglucosyl groups rather than maltotriosyl residues as reported in the literature. Proton NMR, ^{13}CNMR, and Raman spectroscopy have been utilized for the structural determination of pullulan [35, 36]. The presence of α-(1→4) and α-(1→6) glycosidic linkages at 543 and 480 cm^{-1} bands affirms the linkage in the pullulan. Its structure is regularly regarded as an intermediate between amylase and dextran structures due to the presence of these two linkages in the single compound.

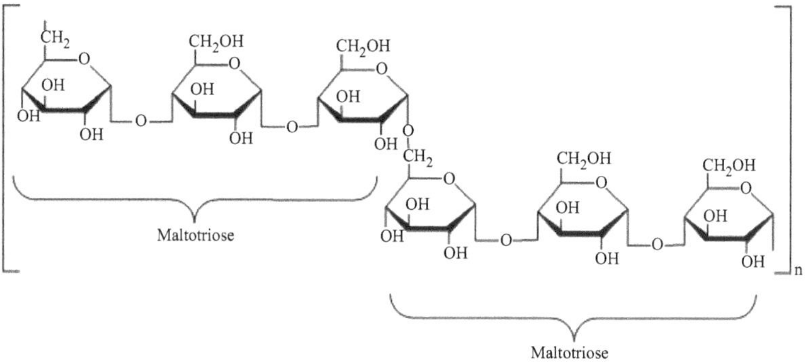

Figure 5.1 Chemical structure of pullulan.

5.3 Properties of Pullulan

The distinction in the controlled system for pullulan elaboration in the biosynthetic pathways or cell morphology could be the possible reason for the same. It is water-dissolvable yet insoluble in natural and non-hygroscopic solvents. The aqueous solution of pullulan is stable and demonstrates a moderately low thickness when contrasted with different polysaccharides. Hydrogels (film forms) are transparent, thermostable, versatile, and antistatic with amazing low-oxygen permeability. It is biocompatible and biodegradable. It decomposes at 250–280°C and is spinnable, malleable, and also acts as good adhesive and binder [37]. Degradation of pullulan results into nontoxic monomers [38]. It is stable in thermal conditions [39]. The main chain of the polymer has a higher number of hydroxyl groups due to which it possesses some inherent physiological activity [40]. The consecutive maltotriose units of pullulan having α-(1→6) linkages are thought to provide enhanced solubility and structural flexibility [41]. Similar reports have also been confirmed by Forabosco et al. [42]. This enables pullulan to imitate synthetic polymers extracted from petrochemical polymers in many ways, such as biodegradability, biocompatibility, and compatibility between humans and the environment. Except dimethylformamide and dimethyl sulfoxide, it is insoluble in natural solvents such as ethanol and isopropanol [43]. The aqueous solutions of pullulan are viscous but do not create gels. The viscosity of pullulan water solutions is proportional to the molecular weight of the pullulan. It is also stable in solution form over a broad spectrum of pH circumstances. Even when burned, it releases no hazardous gas and is spontaneously decomposed by microorganisms even when discarded. Pullulan has great adhesive characteristics when dried and great retention characteristics of foam when dissolved in water. Together with its low viscosity, the exceptional adhesive characteristics of pullulan allow active ingredients to be applied efficiently to the body. This also makes it appropriate as a binder for agglomerated products, helping to enhance the efficiency of current manufacturing procedures by giving the consumer the chance to generate fresh product presentations. Pullulan also has excellent oxygen barriers and is designed for food packaging [44].

5.3.1 Physical Properties of Pullulan

Pullulan is a white or yellowish-white powder with an average molecular weight (M_n) of about 100–200 kDa and a weight-average molecular weight (M_w) of about 362–480 kDa. It is edible [45], flavorless, odorless, tasteless, nonmutagenic, and nontoxic [46]. Values of M_w/M_n were frequently recorded and ranged from 2.1 to 4.1 [47–49]. These values are significantly smaller compared to other exopolysaccharides of industrial importance such as amylose and dextran. Pullulan is an adhesive and demonstrates superior adhesivity on wood compared to corn starch and modified cellulose, when dissolved in water. Pullulan-based films are edible, clear, impermeable in oil, and easily soluble in water [50, 51]. When dissolved in warm or cold water, pullulan forms stable, viscous solution but does not create a gel. It is insoluble in all natural solvents with the exception of dimethylformamide and dimethyl sulfoxide [52, 53]. It may become insoluble in water and soluble in acetone or ethanol when compared to ether and ester, depending on the degree of substitution. The viscosity of pullulan is less than that of other hydrocolloids. The viscosity of pullulan solution is not impacted for a broad range of pHs and remains stable in the presence of ions and is resistant to heat also [54]. Certain ions, such as borate, form a hydroxy pullulan complex that increases viscosity. However, even those solutions do not form gels [55].

Thermal stability is increased by hydrogenation; reaction of hydrogen gas with pullulan aqueous solution results into hydrogenated pullulan in the presence of catalysts under pressure until the disappearance of the reducing property of the solution [23]. Upon carboxylation, the solubility of pullulan is enhanced in cold water [56]. Carboxylated pullulan produced through an oxidizing pullulan has favorable film-forming characteristics, which lead to a film that possesses high mechanical strength, high elasticity modulus, and excellent transparency [57].

5.3.2 Chemical Properties of Pullulan

Pullulan is a neutral polysaccharide. Modifications can result in incorporation of anionic charge as in pullulan sulfate, pullulan carboxylate or a negative charge as in alkylamino pullulan and a high molecular weight than native pullulan [57–59]. In iodine solution, pullulan shows no color of reaction [15, 14, 60]. An insoluble complex of blue copper is produced from one study; no such reactions have been found by any other studies. The pre- and post-hydrolysis ninhydrin test produces negative outcomes. Only carbon dioxide and mild heat are released during burning of pullulan, and combustion heat is about 16.7 kJ/g (4 kcal/g). Pullulan periodate oxidation produces formic acid as a by-product. The ratio of (1→6)-α-D- to (1→4)-α-D-glycosidic linkages can be calculated theoretically from the amounts of formic acid produced and periodate consumed. Permethylation can also determine the ratio of linkage [61, 62]. Complete hydrolysis of acid (e.g., 1 N HCl at 100°C for 120 min) is a by-product of glucose. Partial acid hydrolysis produces different mono- and oligosaccharides, depending on hydrolysis circumstances and processes.

5.3.3 Biochemical Properties

Hydrolyzing enzymes of pullulan, more commonly referred to as debranching enzymes for other polysaccharides, are of great importance. Pullulan is hydrolyzed enzymatically by both (1→6)-α-D- and (1→4)-α-D-pullulanases.

5.4 Applications of Pullulan

5.4.1 Pullulan in Cosmetics

Pullulan is used as additive and thickener in cosmetic formulations for providing smoothness due to its stable viscous nature in a wide range of pH, heat, and in most metal ion conditions. In cosmetics, pullulan gives an instant tightening effect by creating a film that stays on the surface of the skin. This effect is purely cosmetic and washes away after use, and it enhances the immune response and makes the skin more resistant to infections. Pullulan has exceptional transparent film-forming ability, water solubility, moisture absorptivity, and tackiness, which make pullulan appropriate for use in cosmetics. Pullulan typically has low viscosity in aqueous solution than any such high polymer for cosmetics. Pullulan, being non-irritant and nontoxic to the human body, may be useful to any cosmetics but is favorably used as a constituent of cosmetic lotions, cosmetics around eyes, cosmetic powders, shampoos, facial packs, specific hair dressings (hair lacquers and set lotions), and tooth powders (Fig. 5.2). Leung et al. revealed physiologically acceptable films with edible films prepared from pullulan [63]. These edible films include pullulan and antimicrobially active quantities of essential oils such as thymol, methyl salicylate, eucalyptol, and menthol, and these films are effective killers of plaque-producing gums that cause dental plaque, gingivitis, and bad breath [63]. Nakashio et al. investigated the use of pullulan in cosmetics, lotions, and shampoos [64].

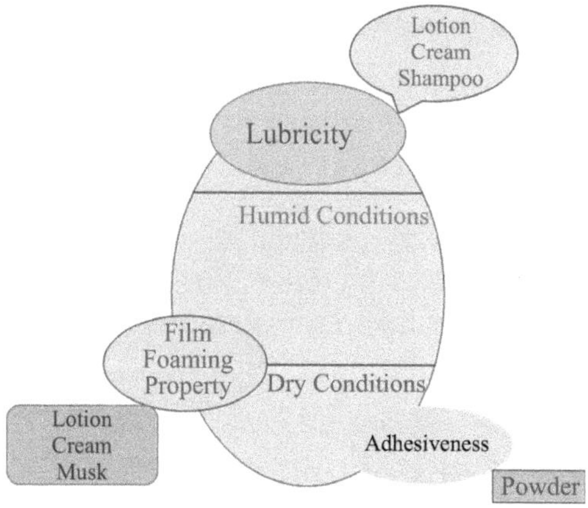

Figure 5.2 Application of pullulan in cosmetics.

5.4.2 Food-Related Applications of Pullulan

The potential applications of pullulan in the food industry are due to its unique structural properties. The odorless, tasteless, and biodegradable properties allow pullulan in food supplements as non-polluting wrapping material [1]. The low calories production and GRAS (generally regarded as safe) grade allow pullulan to be treated as dietary fiber for humans and rats, resulting its direct protection from mammalian amylases [65–67]. Studies demonstrated that the dietary pullulan works as a prebiotic [68] and advances the development of valuable *Bifidobacterium* [69]. Pullulan can be introduced to replace starch from both solid and liquid food supplements, providing properties such as moisture retention, dispersibility, and consistency to foods usually extracted from starch [70]. The moisture content in food is preserved by films, resulting in microbial growth reduction by inhibiting the respiration of microbial flora. Several nations have authorized to commercialize pullulan-based products such as mouth freshners and other oral care products under the brand name "Listerine." [11]. Pullulan may also be used in pastas or baked products as a partial substitute for starch. Pullulan increases the food's shelf life as it is not an easily assimilable source of carbon for food spoilage caused by bacteria, molds, and fungi. In water retention, pullulan is also superior to starch, thus preventing food spoilage by drying out [70–72].

Pullulan can be used in sauces and beverages as filler due to its low viscosity like gum arabic. It is used as a tobacco binder [73], as fertilizers for plants, and as seed coatings [74]. Pullulan and its derivatives also display adhesive characteristics [75]. It can adhere nuts to cookies. It can be used in food pastes as a stabilizer, emulsifier, and binder. The polymer possesses lubricant and gelling abilities. The quality and texture of mayonnaise may also be stabilized by pullulan polysaccharide. Factors such as pH shifts, heating, and most metal ions, including NaCl, do not affect the pullulan viscosity. Pullulan is used as glazing agent in various foods. Cholesterol-substituted pullulan or fatty acids can be used to stabilize fatty emulsions, concluding the use of pullulan as slowly digested carbohydrate and its introduction into food products, in particular drinks, and replacement meal products [76, 77]. The slowly digested carbohydrate pullulan polymer is readily converted into glucose by human enzymes resulting in increasing the glucose level in human blood. For people with diabetes, dietary snack foods have been designed in which pullulan has been integrated. Pullulan also benefits patients with impaired tolerance of glucose.

Films of pullulan displayed optical clarity, high oxygen barrier, and good mechanical flexibility. Thin pullulan films (5–60 μm) are developed by drying a solution of pullulan (regularly 5–10%) on smooth surfaces [78]. The undeveloped films disintegrated in water after a certain period. These films possess the property of melting in mouth as edible food coatings [79]. The oxygen-proof pullulan films protect fats from oxidizing and prevent the vitamin-containing foods from degradation. The pullulan-based films can be employed as food coating and packaging materials of dried foods such as meats, vegetables, noodles, nuts, and confectionaries [80].

5.4.3 Biomedical Applications of Pullulan

Pullulan is currently widely examined for different applications in the biomedical field. This is, for the most part, because of its noncarcinogenic, nontoxic, nonimmunogenic, nonmutagenic, and biodegradable properties. In contrast to a comparative yet progressively well-known polysaccharide, dextran, the rate of degradation of pullulan in serum is quicker than that of dextran. After the incubation period of 48 h, 0.7 is the degradation index in correlation with 0.05 for dextran. The rate of degradation can be decreased or managed by differing degrees of synthetic modification [81]. A portion of the significant regions wherein pullulan is examined are discussed about in the accompanying area. Derivatization of pullulan is a simple procedure to upgrade its action and to enlarge its applications in the biomedical field. Pullulan derivatives have been obtained by graft copolymerization with diverse chemical moieties on its backbone [82]. The hydroxyl groups present on the pyranose ring of pullulan offer many sites for derivatization. Distribution of the hydroxyl groups gives the best possible appearance to the compound. For every repeating unit, pullulan has nine hydroxyl groups; it tends to be derivatized in different structures by substituting these hydroxyl groups to upgrade its utility in biomedical applications. The derivatization differs as indicated by the solvent and reagent nature [24]. Different derivatives are synthesized through various reactions such as copolymerization, esterification, chlorination, etherification, sulfation, amidification, and oxidation [83].

5.4.3.1 Pullulan in drug-delivery systems

Pullulan can be formed in distinct structures, such as microspheres, nanoparticles, or nanogels, which have been accounted for in the ability to deliver drugs. These nanoparticles experience internalization in the cell through an integrin-intervened connection [84. 85]. Integrins are used as cell-attachment receptors for an association between cells and nanoparticles. This link initiates focal adhesion kinase enzyme that helps to revamp cytoskeletons, prompts endocytosis [86, 87], and leads to the internalization of nanoparticles [88].

The methodology of targeting the drug can be classified into active and passive targeting. Modification of surface particles with targeting materials occurs in active targeting. Binding of antibodies or other adhesives to the particle surface follows a proper bio-specific identification mechanism. The recognition of the surface of cells is additionally changed with the counterparts like antigens for specific drug delivery to the site of activity pursued by potential intracellular take-up [89, 90]. Passive targeting can be achieved through multiple organizational courses or after intravenous implementation or toward an epithelium that does not alter the surface of the particles. The reticuloendothelial structure (liver or spleen) perceives particles with a hydrophobic surface that is essentially regulated [91]. These particles are then absorbed by macrophages through phagocytosis and allow a specific collection of drug-delivery systems based on nanocarrier in the perfect region [92].

Pullulan nanoparticles, nanogels, and microspheres can function as effective drug-delivery systems with an increased impact on permeability and retention. Drugs combined with these types tend to accumulate much more in tumor/diseased tissue or cells than in ordinary tissues and decrease drug toxicity to ordinary cells. With the assistance of cell adhesion receptor integrin [84] such as lectin-like receptors on liver cells, these forms can undergo cell internalization, which leads in the greater biological affinity of pullulan to the liver [93, 94]. Pullulan has a brief life span in the blood circulation in comparison to dextran and is accumulated in the liver owing to its high molecular weight [95, 96]. Interferon gets conjugated by pullulan, and this complex helps manage the disease of the hepatitis C virus by targeting the liver directly without any side effects [97, 38]. When injected intravenously, it can also control the demand for the liver-targeting dose. In the pullulan structure, the orientation of hydroxyl groups on the rings of pyranose helps to properly bind pullulan to asialoglycoprotein receptors, which further help in endocytosis.

Incorporating cyclodextrin into pullulan microspheres enhances drug (diclofenac) stability, dissolution rate, and bioavailability [98–101]. These can be polymerized or connected to preexisting materials using a crosslinking agent such as diisocyanates or epichlorohydrin [102–104]. Crosslinked pullulan microspheres with cyclodextrins (α, β, or γ) use epichlorohydrin as a crosslinked agent in the proportion of 1:1 (cyclodextrin: epichlorohydrin). Further diclofenac attached to cyclodextrin-modified pullulan microspheres was used as a model drug [98]. By the addition of the required oligomer pendants to neutral pullulan, several pH-sensitive and temperature-dependent drug-delivery systems can be prepared. The pullulan microspheres with oligomer pendants of semitelechelic poly(N-isopropyl acrylamide-co-acrylamide) function as temperature-dependent and nontoxic drug-delivery structures. It has been observed that pullulan microspheres with extended thermosensitive pendant units are nontoxic at concentrations of up to 10 mg/mL [105]. The on–off switching behavior of these oligomer pendants with respect to lower critical solution temperature (LCST) shows the nature of drug release, i.e., below LCST pendants remain in an extended form to limit the release of drugs and above LCST pendants decrease to allow drug delivery [106–108]. Likewise, in normal cells and cancer cells, pH-sensitive pullulan drug-delivery systems operate based on a pH difference. The pullulan–doxorubicin pH-sensitive conjugate is stable only at neutral pH. However, there is a reduced pH in the tumor cells, resulting in drug release in tumor cells. This delivery system is secure, leads to no hemolysis, time period of longer circulation in blood, and has increased drug retention in low cardiotoxicity tumor cells as well [109, 110]. Doxorubicin release was compared by using Gantrez-PLGA (poly(lactic-co-glycolic) acid), Gantrez-PES (poly(lactic-co-glycolic) acid), and Gantrez A 119 ISP-pullulan nanoparticles. By using doxorubicin as a drug, hepatic cancer was treated with pullulan functionalized nanoparticles via receptors of asialoglycoprotein. These receptors hold the drug, trap the sugars, and help in sugar internalization [111]. Functionalized nanoparticles with asialoglycoprotein ligand receptors are used for targeted drug delivery [112]. The highest drug-loading capacity and entrapment efficiency have been seen in Gantrez–doxorubicin–pullulan nanoparticles.

Synthesis of hydrophobically modified cholesterol-bearing pullulan (CHP) is carried out by using alkylenediamine and sodium monochloroacetate in the presence of 1-ethyl-3(3-dimethylamino)-propyl carbodiimide hydrochloride for binding the cholesterol group to the main chain of pullulan. For hydrophobic molecules, pullulan acts as a good carrier [113, 114]. Accumulation of CHP occurs in aqueous solutions, forming spherical hydrogel nanoparticles [115–117]. At a concentration of 2%, the solution starts to become viscous, and at a concentration of >3.5%, self-accumulation of the polymer occurs [118]. The hydrophobically modified CHP also acts as a carrier for insulin. Insulin as a carrier additionally helps in bringing down the unnecessary blood glucose level (50–60%) in diabetic patients [119]. Conjugation of CHP–insulin shows greater biological activity and long-term stability than native insulin. It can decrease the negative impact of drug frequency injections and constant injections. Only a low drug dose is effective for the proper delivery of drugs [120, 121]. Nanogels of CHP could provide a valid supplementary strategy to neurological disorders with antibody immunotherapy identified by the formation of soluble toxic aggregates, for example, those in Alzheimer's disease [122]. CHP neutral nanogels can communicate with both monomeric and oligomeric A beta-1–42 and diminish the toxicity of beta-1–42 in both essential cortical and microglial cells of the sensory system. Plaque acts as a reservoir for toxic oligomer species; development of plaque with beta-amyloid peptide oligomers in Alzheimer's disease happens due to intra- and extracellular protein aggregation [123]. Nanoparticles assist in the flow of drugs through the blood–brain barrier to the targeted neuron cells [124].

CHP nanoparticles modified by amine result into greater efficiency *in vivo* and *in vitro* as an anticancer drug carrier (docetaxel) for targeting cancer cells in lungs [125]. Both *in vivo* and *in vitro*, CHP-coated liposomes also have great liver-targeting capabilities [126]. The safe delivery from hydrophobically modified CMP nanoparticles to tumor cells occurs via hydrazone bonds with drugs. Antioxidant properties of CMP have strong attraction toward spleen and lymph nodes [127]. PH-sensitive CMP nanoparticles possess property of liver targeting and show well-behavior on different PH [128, 129]. CMP–drug conjugation leads to longer blood circulation of the drug, greater retention in tumor cells, having low cardiotoxicity with no hemolysis [130]. CMP can combine to treat autoimmune diseases with immune suppressants [131]. By adding jeffamines, thermosensitive CMP nanoparticles are created [132]. These nanoparticles are amphiphilic in nature and may maintain acidic, hydrophobic, and fundamental drugs for regulated applications in drug delivery [133]. CMP-Sialyl Lewis X complex can be used to treat various human illnesses (graft rejection, atherosclerosis, ischemia, asthma, etc.). It can readily interact with the expression of the cell-binding molecule E-selectin at inflammatory locations, which enables the delivery of cell-specific drugs in the receptor mediated.

The thermodynamically stable PA micelles of small size (<100 nm) can be used for longer blood circulation of drugs and their targeted delivery to the tissues/tumor cells [134]. PA type of nanoparticles encapsulates hydrophobic drugs in the core, and diseases such as tumor and panic disorder can be easily handled. By adding oligosulfadimethoxine or sulfadimethoxine, nanoparticles can also achieve the property of being pH sensitive to improve the delivery of drugs [135]. The diseased cells/organs with a lower pH can be treated by these nanoparticles than normal blood and cells, such as tumor, cancer, ischemia, etc. PA with carboxymethylated poly(ethylene glycol) increases the release of hydrophobic drugs, as well as helps to prevent the absorption of macrophages and proteins [136, 137]. PA nanoparticles having ionic resistance sensitivity can aggregate with greater ionic resistance (0.15) in the human body and accumulate in tumor cells for targeted radioisotope delivery [138]. PA nanoparticles with folate enhance drug entanglement and effectiveness of discharge. Folate-modified PA exhibits greater cytotoxicity to human nasopharyngeal epidermal cancer cells and improves drug cell absorption [139].

Pullulan production with poly(N-isopropyl acrylamide-co-acrylamide) followed by succinic anhydride treatment results in temperature and pH sensitivity of SP microspheres [108, 140].

The use of diethylenetriaminepentaacetic acid (DTPA) pullulan as an interferon carrier can cure various never-ending liver diseases [141]. By acting as a chelating agent, DTPA decreases the loss of interferon biological activity [142]. For acidic cancer cells, it has greater cellular uptake of the drug and more cytotoxicity.

Pullulan-g-poly(DL-lactide-co-glycolide) and pullulan-g-poly(L-lactide) contribute to cancer cell degradation through controlled and long-term delivery of drugs [143]. L-lactide functions as hydrophobic mediety and can carry anticancer drugs that are water insoluble. Temperature causes the internalization of drug nanogels in cancer cells that are thermosensitive to pullulan-g-poly(L-lactide). These nanogels also have a greater cytotoxic impact on cancer cells and assist in chronic disease therapy [144]. For an effective drug carrier, nanogels of pullulan–deoxycholic acid are used because of having low critical concentration and hydrophobic behavior of the resulting complex [145]. On the apical membrane of polarized epithelia, folate receptors are present, so they have very restricted access to normal cells compared to diseased cells. With the rise in cancer cells, the density of folate receptors rises. Folate pullulan is adorned with an elevated drug therapeutic index and can readily target lung, brain, breast, kidney, and ovary cancer [146]. The pH-sensitive folate-decorated maleilated pullulan (FAMP) improves drug cell absorption, improves cytotoxicity, and has increased antitumor impacts [147]. The nanogels of stimulus-sensitive spiropyrane-modified pullulan display light-regulated drug release as well as tissue-engineering applications [148]. Spiropyranes are a group of "spiroform," a molecule that is colorless, light-switchable, and absorbs only ultraviolent light radiations. It shifts into merocyanine under UV or thermal stimulation when spiropyrane undergoes isomerization. It returns to spiropyrane when the Mer molecule is exposed to visible light. This reversible photochemical method may be biocompatible because there is no generation of free radicals. The delivery of precise drugs by spiropyrane-based nanogels of pullulan provides a particular patient through adjustments in wavelength [149, 150]. A two-stage method with pH- and temperature-sensitive characteristics has recently acquired a hybrid pullulan hydrogel. The crosslinking polymerization of N-isopropyl acrylamide in the presence of carboxymethyl pullulan and the pullulan's secondary reticulation lead to the creation of highly porous structure of hydrogels that improve the swelling/deswelling characteristics and increase the ability of drug loading. A promising candidate as a carrier for drugs against hepatocellular carcinoma is the novel pH-responsive nanoparticle system with reversible pullulan-based shell, poly(β-amino ester) and poly(lactic-co-glycolic acid) core [151]. The third most prevalent cause of cancer death in humans, especially in Southeast Asia and Africa, is hepatocellular carcinoma [152].

5.4.3.2 Pullulan in gene-delivery systems

The derivatives of pullulan can be used as carriers for gene or DNA plasmid to different organs or cancer cells, much the same as delivery of drugs. They can also be used to encapsulate gene or DNA in them in the form of nanogels or nanoparticles [153]. An extensive exploration of application of pullulan in gene delivery has been carried out. Virus-based gene therapy has been conducted, but the significant disadvantage of using viruses is that they are disease-causing, immunogenic, and health hazards. For the implementation of gene delivery, pullulan is explored as biocompatible and nontoxic. Hydrogels of pullulan nanoparticles have an inner hydrophilic core for targeted gene delivery without cytotoxicity [86]. The encapsulation of pBUDLacZ with high loading efficiency in pullulan hydrogel nanoparticles results in a continuous release of DNA. These nanoparticles tend to shield the DNA or gene from the degradation of DNase and have a high potential for transfection. It is possible to deliver targeted genes by binding ligand to the surface of nanoparticles. The diseases of immune system can be treated by immunotherapy [154]. Modification of immunotherapy is carried out by combining pullulan with allergens (CS-560) to treat Japanese pollinosis cedar patients effectively. T-helper cells and cells of mast (Th2, type 2) increase seasonally in number in Japanese pollinosis cedar. The conjugation of allergic pullulan leads from Th2 to Th1 cytokine profile in immune deviation. The various hereditary disorders and cancer-related diseases are believed to be cured by gene therapy. To achieve efficiency and specificity, genes are delivered via endocytic routes. Efforts have been made by using virus for gene therapy, but it is known that viruses are immunogenic and can be dangerous. To this end, attempts are made to develop vectors that are non-viral and cationic natural polymer derivatives. For the implementation of delivery of genes, pullulan has been explored for the implementation of gene delivery being its nontoxic and biocompatible nature. The derivatives of pullulan and plasmid DNA with Zn^{2+} ions having coordination no 17 can be obtained from the derivatives of pullulan–DTPA, which possess chelating metal residues and aqueous solution of plasmid DNA. Pullulan is known for its liver specificity and is used for liver targeting. The authors noted that the conjugate increased gene level expression in the paranchymal liver cells, and increased gene expression kept going 12 days after intravenous injection. Several derivatives of pullulan have been prepared by incorporating diethylenetriaminepentaacetic acid (DTPA), triethylenetetramine (Ti), and spermine (Sm) in pullulan. Through metal coordination, conjugation of DNA with pullulan derivatives occurs to deliver specific DNA [155]. Conjugation of pullulan–spermine complex with plasmid DNA transfected rat brain and endothelial cells of human brain. Complex of pullulan–spermine helps to easily transfect plasmid DNA consisting of cDNA encoding for human growth hormone 1 (hGH1). The mechanism of pullulan–spermine delivery assists brain endothelial cells in the delivery of DNA and protein [156]. The conjugation of pullulan–spermine complex had delivered genes to T24 cancer cells of human bladder. The complex of pullulan–spermine–DNA leads with acceptable cytotoxicity to higher gene transfection and higher cellular uptake. The pullulan–spermine complex can be used through reverse transfection method to treat Parkinson's disease with cell-mediated therapy [157]. Pullulan–spermine helps to move the notch intracellular domain (NICD) gene to stromal cells of the bone marrow (BMSCs), which possesses dopamine-generating neuronal cells that secrete dopamine for the therapy of Parkinson's disease through inverse transfection techniques [158].

For gene transfer to local arteries or vascular smooth muscle cells, hydrogels of tubular DEAE pullulan may be used. The excellent cationic nature of these hydrogels makes them easier for the gene to target the required site. Hydrogels of DEAE pullulan can be molded into three-dimensional matrices to prevent secreted embryonic alkaline phosphatase (SEAP) plasmid from DNase degradation or the protection for SEAP genes [159]. DEAE pullulan can be used for gene therapy by delivering small arterial interfering RNA (siRNA) and facilitates the silencing of genes in vascular cells. In the presence of 2-methacryloxyethyl phosphorylcholine (MPC), cholesterol-bearing pullulan (CHP) complexed with methacryloyl group via radical polymerization has chaperone-like property. Nanogels carrying methacryloyl group pullulan (CHPMA) can readily load enzymes (carbonic anhydrase B) or proteins (insulin) to avoid denaturation and aggregation. Even proteins or denatured enzymes result in refolding through guest–host interaction when trapped in by CHPMA nanogels [160, 161]. Therefore, CHPMA enables to improve the heat stability of the gene or protein for further shipment to specific locations in pullulan as a therapeutic tool in biomedical applications. It has been shown that the unfolded form of carbonic anhydrase B denatured by heat becomes more thermostable when complexed with CHP. The addition of β-cyclodextrin results in the discharge of unfolded CAB from complex and the refolding of CAB to give its indigenous form [162]. Complexes of polyethyleneimine (PEI) pullulan–siRNA are used to treat diseases that are incurable, such as viral infections, tumors, etc. [163, 164]. Mice were injected with the conjugate of siRNA–PEI–pullulan through the tail vein. The delivery of liver-specific siRNA was done with the support of PEI pullulan. It seems to be more stable in the presence of plasma, compatible with blood, and allows delivery of targeted genes to the liver [165]. Pullulan–glycidyl trimethyl ammonium chloride offers excellent hepatic targeted transfection effectiveness, gene stability, and blood compatibility [166].

5.4.3.3 Pullulan in cell/tissue engineering

Tissue engineering needs a framework or artificial matrix that can accommodate cells and control their development leading to the regeneration of three-dimensional tissue. Conjugation of carboxymethyl pullulan with heparin leads to examination of its characteristics toward tissue-engineering applications [167]. The proliferation of smooth muscle cells (SMCs) was inhibited by heparin-conjugated pullulan. It is, therefore, possible to use heparin-conjugated pullulan material for the proliferation of endothelial vascular cells and inhibition of SMC proliferation. The cell-appropriate atmosphere is developed by using a suitable three-dimensional artificial scaffold for damaged organs or tissues [168, 169]. The self-healing capacity of these damaged tissues is improved by tissue engineering. The design of three-dimensional scaffolds differs with the form of extracellular matrix (ECM) collagen fibers of damaged tissue [170].

Various biopolymers with indistinguishable properties from natural tissues can be shaped into various structures, for example, frameworks, hydrogels, smaller-scale grids, miniaturized scale dots, and tissue-engineering applications, scaffolds, micro molded matrices, hydrogels and microbeads [171]. The usage of these biopolymers basically incorporates surface modification techniques in tissue engineering. Pullulan's surface characteristics can be readily improved by replacing the required chemical moieties on its hydroxyl groups [172]. Pullulan has remarkable mechanical qualities such as incredible biocompatibility and the high capacity for hydration [173, 174]. As a result of these qualities, scaffolds of pullulan assume a promising job in empowering cell-based dermal substitution, vascular cell tissue engineering, and bone regeneration. Scaffolds of pullulan hydrogel help to build complex of cell-laden micro-tissue to encapsulate damaged organs or cells in order to regenerate and proliferate them [75]. The hydrogels have atmosphere hydrated, pore composition interconnected, and macroporosity. They help the targeted sites to release regulated metabolites and nutrients [175]. In maxillofacial and orthopedic surgeries, pullulan/dextran composite hydrogels are applied. They assist to stimulate host mesenchymal stem cells (MSCs) differentiation of bone cells. In addition, the coating of nanocrystalline hydroxyapatite particles (nHAP) enhances the mechanical characteristics of these three-dimensional hydrogels and also enhances the capacity of cells to attach to them. The coating of three-dimensional structures of pullulan helps in greater osteoconductivity and enhances their compressive modulus. The pullulan/dextran scaffolds modified by nHAP can activate early calcification, leading to the formation of osteoid tissue in bone and non-osseous locations [176]. New pullulan bioconjugate can be used to treat selective metastases of bone in breast cancer [177]. As a substratum for vascular EC development, heparin-conjugated carboxylated pullulan is suitable [167]. Chondroitin sulfate CMP hydrogels function as a scaffold for the regeneration of cartilage tissue [178]. The pullulan hydrogel system of cholesteryl and acryloyl group-bearing pullulan induces infiltration of osteoprogenitor cells both inside and outside the hydrogels, which in the future help to supply various proteins to defects of skull bone [179].

Hydrogels have received considerable attention as wound dressings due to their physical properties comparable to natural soft tissues and the ability to absorb considerable amount of aqueous liquids. The wound dressings by hydrogels can provide the wound with a humid setting and shield it from bacterial diseases [180–182]. CMP–cystamine-based hydrogels have all these characteristics together with great mechanical strength and can possibly be used as an optimal wound dressing for clinical use [183]. These hydrogels also function as carriers without any reduction in their bioactivity for antimicrobial agents. Consistent porosity of pullulan–collagen composite hydrogels can reproduce dermal architecture and effectively integrate stem cells for early wound healing [184]. Due to their antioxidant characteristics, these hydrogels can protect MSCs from oxidative damage [185]. Modification with salt-induced phase inversion method increases the recruitment and activation of cells in order to heal wounds.

Adding mesenchymal stem cells derived from adipose to pullulan/collagen hydrogels enhances the expression of both transcriptional and protein concentrations of monocyte chemoattractant protein-1 [186]. Adipose-derived stem-cell-seeded hydrogels have potential for increased angiogenesis and continuous recruitment of cells. The excellent adhesion property of phosphorylated pullulan (PPL) with hard tissues occurs through ionic bonding and helps to repair bone defects [187]. PPL can function as an antibacterial agent carrier and bone regeneration enhancer, for example in defects of femur. In addition, the addition of β-tricalcium phosphate improves the biocompatibility, water solubility, and mechanical characteristics of PPL for bone-tissue engineering [188]. Conjunction of carboxylated pullulan with human-like collagen and 1,4-butanediol diglycidyl ether can function as a successful smooth filler for tissue engineering [189]. For skin-tissue engineering, the three-dimensional pullulan–cellulose acetate scaffolds have cytocompatibility to facilitate cell attachment, spread, and proliferation [190].

CHP nanogels can capture tissue-engineering hydrophobic medicines. Conjugation of nanogel CHP with prostaglandin E1 creates a humid atmosphere in complete thickness wounds that promote neoepithelialization, neovascularization, and regulated release of prostaglandin E1 for injury regeneration [191]. The novel pullulan amphiphilic nanogel-crosslinked hydrogels have powerful mechanical characteristics, greater biodegradability, and can act as universal tissue-engineering scaffolds [192]. Even without growth factors, the hydrogels can function properly to enhance infiltration of cells and regeneration of damaged cells. The pullulan/dextran-nHAP microbeads can function as prospective scaffolds to repair variable size bone defects [193]. These microbeads can increase the development of osteoid tissue by improving cell interconnectivity and maintaining their structural trabecular structure. Another useful way to repair defect of bones is pullulan microspheres [194]. The modified surface of these microspheres has injectable size biocompatible spherical microcarriers that can be applied to tiny incisions. By adding trisodium trimetaphosphate, simulated body fluid incubation and silk fibroin coating, the compatibility and stability of pullulan microspheres can be improved.

5.4.3.4 Pullulan in antibacterial release for wound-dressing applications

In recent years, antibacterial wound dressings have become increasingly common, with most commercial providers now providing dressings impregnated with silver or nanoparticles [195, 196]. While the extensive spectrum of antibacterial effectiveness of such dressings has been shown, several studies have also reported on silver cytotoxicity to mammalian cells and an adverse effect on the wound-healing process [197, 198]. Besides, silver wound dressings are costly, and there are continuous concerns over absorption and nanotoxicity of nanoparticles. Therefore, considerable effort has been made to identify alternative antibacterial agents [199]. TA/chitosan/pullulan composite nanofibers were synthesized for wound-healing applications, although these materials were not pH responsive [200]. Pullulan-based hydrogels were synthesized by chemical modification in order to achieve perfect wound dressing. In aqueous fluids, the absorbing capacity of hydrogels showed the ratio of swelling up to 4000%, which furnish the hydrogels with excellent ability of rapid hemostatic capacity and prevented the wound bed from accumulating exudates. As fresh perfect wound-dressing materials, the pullulan-based hydrophilic hydrogels have high potential. These hydrogels can retain a humid media over wound bed, which could stopover the wound bed dehydration and prevents the formation of scabs. The pullulan-based hydrogels are biocompatible with no cytotoxicity. Wounds can be protected from the invasion of bacteria and bacterial growth could be prevented by loading antimicrobial agents such as gentamycin sulfate in hydrogels. The natural pullulan-based hydrogels have outstanding properties of wound dressing and may be guaranteed in clinical applications [183]. The bacterial infections due to surgical wounds and pediatric skin burns are primarily caused by *Escherichia coli* and *Staphylococcus aureus* strains. Pullulan/peptide-based nanofibers could be regarded as an administration system in skin burn therapy for the design of wound dressing. The incorporation of antimicrobial LfcinB-derived peptides to the pullulan nanofibers and the pullulan/peptide nanofibers liberates the peptide in instantaneous manner. Pullulan–LfcinB (21–25) Pal nanofibers could be considered for designing and developing antibacterial wound dressings [201]. An organic disulfide agent (cystamine) having covalent crosslinking may adjust the physical characteristics such as swelling ratio and tensile strength of the pullulan-based hydrogels for the desired release of antibacterial agents [202]. Chitosan/pullulan composite nanofibers have been developed for wound-healing applications and antibacterial activity against *E. coli*, a Gram-negative bacteria, although these materials are not pH sensitive [200].

5.4.3.5 Anticancer applications of pullulan

Pullulan is a versatile molecule designed to functionalize chemical changes and to build the opportunity of electrostatic type of hydrogen bonding or hydrophobic interactions with other small-molecule compounds. In some instances, pullulan itself plays the role of a cell target, such as receptors of liver or cancer cells. The derivatives of pullulan can be conjugated or may be complexed with hydrophobic drugs. Scientists optimize hydrophilic–hydrophobic equilibrium and physical interactions to generate the drug-loading controlled systems and drug delivery to receptors of cancer and liver cells. The demand for anticancer drugs has been growing for the previous centuries; several nontoxic, biocompatible, and biodegradable systems based on pullulan derivatives have been tested and designed for loading and release of 10-hydroxycamptothecin or mitoxantrone, doxorubicin, paclitaxel, and epirubicin. Pullulan-based anticancer drug-delivery systems include hydrophobic pullulan self-assembly or bioconjugates and pH-sensitive pullulan nanoparticles. Conjugation of bioactive drugs with pullulan as bioconjugates was intended to produce an acid-sensitive hydrazine bond that is stable under normal physiological pH but hydrolyzes under acidic conditions. There has also been a great deal of effort to find the finest ligand to selectively target drug-delivery systems to a tumor tissue [203]. Because of nontoxic and noncarcinogenic nature of pullulan, it is used as a drug transporter to decrease the toxicity of anticancer drugs to ordinary cells and to increase their therapeutic importance [204, 205]. Lately, all-transretinoic acid (ATRA)-based pullulan (PURA) nanogels have been synthesized by Lee et al. [206]. The impact of anticancer drug is improved with the help of ATRA. It is well-recognized as a therapeutic anticancer agent for various types of cancer, including lung, skin, neck, kidney, and breast cancers [207]. Pullulan-g-poly(L-lactide) (PUPL) is hydrophobic and possesses low critical association concentration. The hydrophobic anticancer drugs are carried by hydrophobic moiety like L-lactide [143]. Thermosensitive pullulan-g-poly(L-lactide) nanogels help in cancer cell degradation through controlled release of drugs. Temperature causes the internalization of nanogel-containing drugs in cancer cells to have a stronger cytotoxic effect [144]. Conjugation of pullulan and deoxycholic acid occurs via ester linkage to develop novel self-organized nanogels for anticancer drug delivery. The self-organized nanogels were observed for their efficacy of drug targeting to cancer cells by conjugating with doxorubicin. The drug loads effectively in these nanogels and possesses low critical association concentration, leading to targeted drug delivery in an appropriate quantity [145]. Epirubicin drug is used to treat throat cancer and has shown greater cytotoxicity to tumor cells than free epirubicin when conjugated with pullulan acetate (PA) nanoparticles [208]. The receptors of folate can act as binding agents when PA nanoparticles get modified with folate. Epirubicin loaded in folate-modified pullulan acetate (FPA), when compared with indigenous PA, has shown greater drug trapping, quicker release of drugs, greater cellular uptake, and greater cytotoxicity to human nasopharyngeal epidermal cells [139].

5.4.3.6 Application of pullulan in bioimaging system

Bioimaging is a comparatively latest development of digital technology in the medical sector. It is, therefore, a strong instrument for visualizing the body's inner functioning and diseases. Of late, nanotechnology is utilized for identifying cancerous cells in the human body. The semiconductor-based nanosized particles known as quantum dots (QDs) have pulled in numerous researchers in the biological field. They are utilized as fluorescent tests for live cell image tracking to specify entire cells or intracellular biomolecules. Semiconductor nanocrystals or QDs have excellent characteristics, such as broad excitation, high photostability, thin spectrum outflow, and bright fluorescence [209–213]. However, delivering QDs to living cells is a significant challenge as there is usually low cell uptake effectiveness for QDs. Because of their separate surface and size characteristics, many QDs are needed for adequate cell imaging. The cationic liposome can function as a carrier for QDs in order to solve the above problems [214, 215]. The cell membranes that are negatively charged get interacted with cationic liposomes in order to promote the endocytosis. Instead of providing QDs with increased effectiveness, several hundred nanometers of aggregates are formed in the cytoplasm from liposomes that limit their usage in bioimaging. Amino group–modified, cholesterol-bearing pullulan nanogels have been developed for the delivery of QDs into cells as a substitute to the regular cationic liposomes, which have the risk of aggregating once they enter the cells [216]. QDs conjugated with CHP possess higher efficiency of cellular uptake in comparison to conventional carriers. QDs-CHPNH2 monodisperse hybrid nanoparticles have promising application for medical imaging of different human cells as a fluorescent probe. Conjugated IRDye-900 pullulan–cholesterol nanoparticles "near-infrared polynanogel" (NIR-PNG) shows the best characteristics of longer retention and least dispersion in the sentinel nodes. These NIR-PNG tracers have excellent sensitivity and specificity for sentinel nodes and can possibly be used for sentinel node navigation surgery as optimal tracers in gastric cancer [217]. DTPA can be used for heptocyte magnetic resonance imaging to treat cancer [218].

5.4.4 Environmental Applications of Pullulan

5.4.4.1 Environmental remediation

Due to the adhesive nature of producing strains, pullulan has heavy metal removal capability. Breierova et al. reported that heavy metals can be removed from aqueous solutions by *A. pullulans* [219]. Radulovic et al. described that peat from Vlasina Lake can be used for *A. pullulans* strain as a natural medium. In the liquid phase, this microorganism can be used for effectively decreasing the metal content, since accumulation of metals with the biomass of *A. pullulans* strain occurs during the fermentation process on acid peat hydrolysate. The key benefit of this method over conventional methods of metal elimination is that using this microorganism concurrently exhibited two valuable effects: production of pullulan and reduction in metal pollution. The advantage of applying this biotechnology for the purpose of reducing metal pollution is because of its dual effects: biosynthesis of pullulan with *A. pullulans* and simultaneous biosorption of metals [220].

5.4.4.2 Filtration and chromatography

Motozato et al., for the first time, used pullulan in the application of gel permeation chromatography as a packing material. They prepared spheres of pullulan by the crosslinked method having an extensive variety of properties, such as high flow rates. They revealed that there were no complications in using pullulan spheres for gel permeation chromatography. The packing in columns displays nearly no interactions (hydrophobic interactions) with sample substrates, and because of its ionic nature, it can be used over a wide range of pH [221]. Fundueanu et al. prepared microspheres of pullulan by chemical crosslinking of cyclodextrin with epichlorohydrin in an alkaline solution of pullulan. Application of pullulan-based microspheres was analyzed in glass column of liquid chromatography by exploring the performance of isocratic elution of various drugs chosen as model drugs. The increase in the retention volume of industrial compounds depends on the interactions between cyclodextrin cavity and the chosen molecule and is explored by the widening of peaks. The precise estimation of the drug can be done by conditioning the potential drug in a suitable cyclodextrin cavity and the values of interaction coefficient can be used to anticipate the release profiles of the drugs considered.

5.5 Challenges and Future Perspectives

Pullulan has gained a lot of attention in the recent years as can be seen from the recent literature. The derivatization of pullulan has been reported to reduce its water solubility, to introduce reactive functional groups, and to provide application-specific derivatives. Hydrophobically modified pullulan derivative nanoparticles are useful in drug-delivery applications. The adhesive nature of pullulan is helpful in wound-dressing applications. In conclusion, it seems that pullulan should be explored toward different applications so that its unique properties can be tapped to its maximum potential.

The precise cell nature accountable for pullulan production is still under examination. Bioreactor proposal regarding cell immobilization and biomass build-up have attracted much attention, which is supposed to improve pullulan production. Novel pullulan separation and purification approaches would be explored, which can decrease the input of labor and money. The natural pullulan and its derivatives own various applications because of their exceptional structural properties. Only *in vitro* studies have been carried out for these chemically improved pullulan derivatives. *In vivo* efficacy of the pullulan derivatives should be investigated. Pullulan hydrogels having non-fouling properties are less explored in bone-tissue engineering applications. Clinical trials should be done to examine their role in helping the development of bone tissues, which leads to bone development in large bone defect. Pullulan-based hydrogels having tubular molds can be investigated for *in vivo* establishment of vascular cells, which may lead to increase in cellular growth and more tissue formation. For bone-healing efficiency, pullulan hydrogels having cholesterol should be studied as a tool for enlightening protein dependence. Pullulan derivatives and their conjugation with proteins may open the option of vaccination and to control tumor expansion by using them in immunotherapy. These findings propose the developing innovations that can enhance the properties of pullulan and their derivatives, thus opening new avenues for applications of pullulan in diverse fields.

References

1. Singh, R. S. and Saini, G. K. (2008). Pullulan-hyperproducing color variant strain of *Aureobasidium pullulans* FB-1 newly isolated from phylloplane of *Ficus* sp. *Bioresour. Technol.*, **99**(9), 3896–3899.
2. Byrom, D. (1991). *Biomaterials: Novel Materials from Biological Sources*, Springer.
3. Garcia-Ochoa, F., Santos, V. E., and Alcon, A. (1995). Xanthan gum production: An unstructured kinetic model. *Enzyme Microb. Technol.*, **17**(3), 206–217.
4. Joris, K. and Vandamme, E. J. (1993). Novel production and application aspects of bacterial cellulose. *Microbiology*, **1**, 27–29.
5. Weiss, R. M. and Ollis, D. F. (1980). Extracellular microbial polysaccharides. I. Substrate, biomass, and product kinetic equations for batch xanthan gum fermentation. *Biotechnol. Bioeng.*, **22**(4), 859–873.
6. Chawla, P. R., Bajaj, I. B., Survase, S. A., and Singhal, R. S. (2009). Microbial cellulose: Fermentative production and applications. *Food Technol. Biotechnol.*, **47** (2).

7. Singh, R. S. and Kaur, N. (2015). Microbial biopolymers for edible film and coating applications. *Adv. Biotechnol. India IK Int. Publ.*, 187–216.

8. Singh, R. S., Kaur, N., and Kennedy, J. F. (2015). Pullulan and pullulan derivatives as promising biomolecules for drug and gene targeting. *Carbohydr. Polym.*, **123**, 190–207.

9. Singh, R. S., Saini, G. K., and Kennedy, J. F. (2008). Pullulan: Microbial sources, production and applications. *Carbohydr. Polym.*, **73**(4), 515–531.

10. Voragen, A. G. J., Pilnik, W., Thibault, J. F., Axelos, M. A. V., and Renard, C. M. G. C. (1995). *Food Polysaccharides and Their Applications*, Marcel Pekker Inc., San Diego.

11. Tsujisaka, Y. and Mitsuhashi, M. (1993). Pullulan. In: *Industrial Gums: Polysaccharides and Their Derivatives* (Whistler, R. L. and BeMiller, J. N., Eds.), Third edn., Academic Press, Inc.

12. Yalpani, M. (2013). *Polysaccharides: Syntheses, Modifications and Structure/Property Relations*, Elsevier.

13. Bauer, R. (1938). Physiology of *Dematium pullulans* de Bary. *Zentralbl Bacteriol Parasitenkd Infekt. Hyg Abt2*, **98**, 133–167.

14. Bernier, B. (1958). The production of polysaccharides by fungi active in the decomposition of wood and forest litter. *Can. J. Microbiol.*, **4**(3), 195–204.

15. Bender, H., Lehmann, J., and Wallenfels, K. (1959). Pullulan, ein extracelluläres Glucan von Pullularia pullulans. *Biochim. Biophys. Acta*, **36**(2), 309–316.

16. Bouveng, H. O., Kiessling, H., Lindberg, B., and McKay, J. (1963). Polysaccharides elaborated by *Pullularia pullulans. Acta Chem. Scand*, **17**, 797–800.

17. Sowa, W., Blackwood, A. C., and Adams, G. A. (1963). Neutral extracellular glucan of *Pullularia pullulans* (de Bary) Berkhout. *Can. J. Chem.*, **41**(9), 2314–2319.

18. Bender, H. and Wallenfels, K. (1961). Investigations on pullulan. II. Specific degradation by means of a bacterial enzyme. *Biochem. Z*, **334**, 79–95.

19. Wallenfels, K. and Bender, H. (1961). *Procedure for the Production of a Dextran-Like Polysaccharide from Pullularia Pullulans.* German Patent Office, Pat. No. 1 096 850.

20. Wallenfels, K., Keilich, G., Bechtler, G., and Freudenberger, D. (1965). Investigations on pullulan. IV. Resolution of structural problems using physical, chemical and enzymatic methods. *Biochem. Z*, **341**, 433–450.

21. Catley, B. J., Ramsay, A., and Servis, C. (1986). Observations on the structure of the fungal extracellular polysaccharide, pullulan. *Carbohydr. Res.*, **153**(1), 79–86.

22. Catley, B. J. (1970). Pullulan, a relationship between molecular weight and fine structure. *FEBS Lett.*, **10**(3), 190–193.

23. Leathers, T. D. (2003). Biotechnological production and applications of pullulan. *Appl. Microbiol. Biotechnol.*, **62**(5–6), 468–473.

24. Shingel, K. I. (2004). Current knowledge on biosynthesis, biological activity, and chemical modification of the exopolysaccharide, pullulan. *Carbohydr. Res.*, **339**(3), 447–460.

25. Choudhury, A. R., Saluja, P., and Prasad, G. S. (2011). Pullulan production by an osmotolerant *Aureobasidium pullulans* RBF-4A3 isolated from flowers of *Caesulia axillaris*. *Carbohydr. Polym.*, **83**(4), 1547–1552.

26. Cheng, K.-C., Demirci, A., Catchmark, J. M., and Puri, V. M. (2010). Modeling of pullulan fermentation by using a color variant strain of *Aureobasidium pullulans*. *J. Food Eng.*, **98**(3), 353–359.

27. Karim, M. R., Lee, H. W., Kim, R., Ji, B. C., Cho, J. W., Son, T. W., Oh, W., and Yeum, J. H. (2009). Preparation and characterization of electrospun pullulan/montmorillonite nanofiber mats in aqueous solution. *Carbohydr. Polym.*, **78**(2), 336–342.

28. Wu, S., Chen, J., and Pan, S. (2012). Optimization of fermentation conditions for the production of pullulan by a new strain of *Aureobasidium pullulans* isolated from sea mud and its characterization. *Carbohydr. Polym.*, **87**(2), 1696–1700.

29. Yuen, S. (1974). Pullulan and its applications. *Process Biochem.*, **22**, 17–22.

30. Chen, L., Ren, F., Zhang, Z., Tong, Q., and Rashed, M. M. A. (2015). Effect of pullulan on the short-term and long-term retrogradation of rice starch. *Carbohydr. Polym.*, **115**, 415–421.

31. Tabasum, S., Noreen, A., Maqsood, M. F., Umar, H., Akram, N., Chatha, S. A. S., and Zia, K. M. (2018). A review on versatile applications of blends and composites of pullulan with natural and synthetic polymers. *Int. J. Biol. Macromol.*, **120**, 603–632.

32. Cheng, K.-C., Demirci, A., and Catchmark, J. M. (2011). Pullulan: Biosynthesis, production, and applications. *Appl. Microbiol. Biotechnol.*, **92**(1), 29.

33. Bender, H. (1961). Untersuchungen an pullulan. *Biochem. Z.*, **334**, 79–95.

34. Carolan, G., Catley, B. J., and McDougal, F. J. (1983). The location of tetrasaccharide units in pullulan. *Carbohydr. Res.*, **114**(2), 237–243.

35. Arnosti, C. and Repeta, D. J. (1995). Nuclear magnetic resonance spectroscopy of pullulan and isomaltose: Complete assignment of chemical shifts. *Starch-Stärke*, **47**(2), 73–75.

36. McIntyre, D. D. and Vogel, H. J. (1993). Structural studies of pullulan by nuclear magnetic resonance spectroscopy. *Starch*, **45**(11), 406–410.

37. Feng, S.-S., Mei, L., Anitha, P., Gan, C. W., and Zhou, W. (2009). Poly (lactide)–vitamin E derivative/montmorillonite nanoparticle formulations for the oral delivery of Docetaxel. *Biomaterials*, **30**(19), 3297–3306.

38. Xi, K., Tabata, Y., Uno, K., Yoshimoto, M., Kishida, T., Sokawa, Y., and Ikada, Y. (1996). Liver targeting of interferon through pullulan conjugation. *Pharm. Res.*, **13**(12), 1846–1850.

39. Barnett, C., Smith, A., Scanlon, B., and Israilides, C. J. (1999). Pullulan production by *Aureobasidium pullulans* growing on hydrolysed potato starch waste. *Carbohydr. Polym.*, **38**(3), 203–209.

40. Singh, R. S., Kaur, N., Rana, V., and Kennedy, J. F. (2017). Pullulan: A novel molecule for biomedical applications. *Carbohydr. Polym.*, **171**, 102–121.

41. Buliga, G. S. and Brant, D. A. (1987). Temperature and molecular weight dependence of the unperturbed dimensions of aqueous pullulan. *Int. J. Biol. Macromol.*, **9**(2), 71–76.

42. Forabosco, A., Bruno, G., Sparapano, L., Liut, G., Marino, D., and Delben, F. (2006). Pullulans produced by strains of *Cryphonectria parasitica*—I. Production and characterisation of the exopolysaccharides. *Carbohydr. Polym.*, **63**(4), 535–544.

43. Sugumaran, K. R., Sindhu, R. V., Sukanya, S., Aiswarya, N., and Ponnusami, V. (2013). Statistical studies on high molecular weight pullulan production in solid state fermentation using jack fruit seed. *Carbohydr. Polym.*, **98**(1), 854–860.

44. Singh, R. S. and Saini, G. K. (2012). Biosynthesis of pullulan and its applications in food and pharmaceutical industry. In: *Microorganisms in Sustainable Agriculture and Biotechnology*, Springer, pp. 509–553.

45. Okada, K., Yoneyama, M., Mandai, T., Aga, H., Sakai, S., and Ichikawa, T. (1990). *Nippon Eiyo Shokuryo Gakkaishi* (in Japanese), **43**, 23–29.

46. Singh, R. S. and Saini, G. K. (2014). Pullulan as therapeutic tool in biomedical applications. *Adv. Ind. Biotechnol.*, 263–291.

47. Wiley, B. J., Ball, D. H., Arcidiacono, S. M., Sousa, S., Mayer, J. M., and Kaplan, D. L. (1993). Control of molecular weight distribution of the biopolymer pullulan produced by *Aureobasidium pullulans*. *J. Environ. Polym. Degrad.*, **1**(1), 3–9.

48. Roukas, T. and Mantzouridou, F. (2001). Effect of the aeration rate on pullulan production and fermentation broth rheological properties in an airlift reactor. *J. Chem. Technol. Biotechnol. Int. Res. Process. Environ. Clean Technol.*, **76**(4), 371–376.

49. Thirumavalavan, K., Manikkadan, T. R., and Dhanasekar, R. (2009). Pullulan production from coconut by-products by *Aureobasidium pullulans*. *African J. Biotechnol.*, **8**(2).

50. Shit, S. C. and Shah, P. M. (2014). Edible polymers: Challenges and opportunities. *J. Polym.*, **2014**.

51. Utami, R., Nurhartadi, E., Yusuf Trisna Putra, A., and Setiawan, A. (2014). The effect of cassava starch-based edible coating enriched with *Kaempferia rotunda* and *Curcuma xanthorrhiza* essential oil on refrigerated patin fillets quality. *Int. Food Res. J.*, **21**(1).

52. Imeson, A. (2010). *Food Stabilisers, Thickeners and Gelling Agents*, Wiley Online Library.

53. Tomasik, P. (2003). *Chemical and Functional Properties of Food Saccharides*, CRC Press.

54. Embuscado, M. E. and Huber, K. C. (2009). *Edible Films and Coatings for Food Applications*, Springer.

55. Stephen, A. M. and Phillips, G. O. (2016). *Food Polysaccharides and Their Applications*, CRC Press.

56. Salatin, S. and Jelvehgari, M. (2017). Natural polysaccharide-based nanoparticles for drug/gene delivery. *Pharm. Sci.*, **23**(2), 84.

57. Tsuji, K., Fujimoto, M., Masuko, F., and Nagase, T. (1978). Carboxylated pullulan and method for producing same. US Patent, US4090016A.

58. Lin, Y. (2010). Improving the production of high molecular weight pullulan using a new radial reciprocating plate impeller. Thesis, University of Ottawa.

59. Mavridis, I. M., Yannakopoulou, K., Eliadou, K., Mourtzis, N., and Aggelidou, C. (2010). Per-6-guanidino-, alkylamino-cyclodextrins, methods of their synthesis and their use for the compaction of DNA and intercellular delivery. US Patent, US7834174.

60. Bouveng, H. O., Kiessling, H., Lindberg, B., and McKay, J. (1962). Polysaccharides elaborated by *Pullularia pullulans*. *Acta Chem. Scand.*, **16**, 615–622.

61. Jackson, E. L. and Hudson, C. S. (1937). Studies on the cleavage of the carbon chain of glycosides by oxidation. A new method for determining ring structures and alpha and beta configurations of glycosides. *J. Am. Chem. Soc.*, **59**(6), 994–1003.

62. Wallenfels, K., Bechtler, G., Kuhn, R., Trischmann, H., and Egge, H. (1963). Permethylierung von oligomeren und polymeren Kohlenhydraten und quantitative Analyse der Spaltungsprodukte. *Angew. Chemie*, **75**(21), 1014–1022.

63. Leung, S.-H. S., Leone, R. S., Kumar, L. D., Kulkarni, N., and Sorg, A. F. (2005). Fast dissolving orally consumable films. US Patent 7025983.

64. Nakashio, S., Tsuji, K., Toyota, N., and Fujita, F. (1976). Novel cosmetics containing pullulan. US Patent, US3972997A.

65. Yatmaz, E. and Turhan, I. (2012). Pullulan production by fermentation and usage in food industry. *GIDA J. Food*, **37**(2), 95–102.

66. Oku, T., Yamada, K., and Hosoya, N. (1979). Effects of pullulan and cellulose on the gastrointestinal tract of rats. *J. Japanese Soc. Food Nutr.*, **32**(4), 235–241.

67. Yoneyama, M., Okada, K., Mandai, T., Aga, H., Sakai, S., and Ichikawa, T. (1990). Effects of pullulan intake in humans. *J. Japanese Soc. Starch Sci.*, **37**(3), 123–127.

68. Mitsuhashi, M., Yoneyama, M., and Sakai, S. (1990). Growth promoting agent for bacteria containing pullulan with or without dextran. European Patent, EP0382355A2.

69. Sugawa-Katayama, Y., Kondou, F., Mandai, T., and Yoneyama, M. (1994). Effects of pullulan, polydextrose and pectin on cecal microflora. *J. Appl. Glycosci.*, **41**(4), 413–418.

70. Hiji, Y. (1986). Method for inhibiting increase in blood sugar content. US Patent, US4629725A.

71. Hijiya, H. and Shiosaka, M. (1975). Process for the preparation of food containing pullulan and amylose. US Patent, US3872228A.

72. Kato, K. and Shiosaka, M. (1975). Food compositions containing pullulan. US Patent, US3875308A.

73. Miyaka, T. (1979). Shaped matters of tobaccos and process for preparing the same. Canadian Patent Office. *Pat*, (1), 49.

74. Matsunaga, H., Fujimura, S., Namioka, H., Tsuji, K., and Watanabe, M. (1977). Fertilizer composition. US Patent, US4045204A.

75. Bae, H., Ahari, A. F., Shin, H., Nichol, J. W., Hutson, C. B., Masaeli, M., Kim, S.-H., Aubin, H., Yamanlar, S., and Khademhosseini, A. (2011). Cell-

laden microengineered pullulan methacrylate hydrogels promote cell proliferation and 3D cluster formation. *Soft Matter*, **7**(5), 1903–1911.

76. Yamaguchi, S. and Sunamoto, J. (1991). Fatty emulsion stabilized by a polysaccharide derivative. European Patent, EP0370810A1.
77. Wolf, B. W. (2005). Use of pullulan as a slowly digested carbohydrate. China Patent, CN1708236A.
78. Prajapati, V. D., Jani, G. K., and Khanda, S. M. (2013). Pullulan: An exopolysaccharide and its various applications. *Carbohydr. Polym.*, **95**(1), 540–549.
79. Kaplan, D. L., Ching, C., and Thomas, E. L. (1993). *Biodegradable Polymers and Packaging*, Technomic Publ. Com.
80. Miller, K. S. and Krochta, J. M. (1997). Oxygen and aroma barrier properties of edible films: A review. *Trends Food Sci. Technol.*, **8**(7), 228–237.
81. Rekha, M. R. and Sharma, C. P. (2007). Pullulan as a promising biomaterial for biomedical applications: A perspective. *Trends Biomater Artif Organs*, **20**(2), 116–121.
82. Hezarkhani, M. and Yilmaz, E. (2019). Pullulan modification via poly (N-vinylimidazole) grafting. *Int. J. Biol. Macromol.*, **123**, 149–156.
83. Popa, V. (2011). *Polysaccharides in Medicinal and Pharmaceutical Applications*, Smithers Rapra.
84. Absolom, D. R., Zingg, W., and Neumann, A. W. (1987). Protein adsorption to polymer particles: role of surface properties. *J. Biomed. Mater. Res.*, **21**(2), 161–171.
85. Haas, T. A. and Plow, E. F. (1994). Integrin-ligarid interactions: A year in review. *Curr. Opin. Cell Biol.*, **6**(5), 656–662.
86. Gupta, M. and Gupta, A. K. (2004). Hydrogel pullulan nanoparticles encapsulating pBUDLacZ plasmid as an efficient gene delivery carrier. *J. Control. Release*, **99**(1), 157–166.
87. Gupta, A. K. and Curtis, A. S. G. (2004). Lactoferrin and ceruloplasmin derivatized superparamagnetic iron oxide nanoparticles for targeting cell surface receptors. *Biomaterials*, **25**(15), 3029–3040.
88. Lyman, S., Gilmore, A., Burridge, K., Gidwitz, S., and White, G. C. (1997). Integrin-mediated activation of focal adhesion kinase is independent of focal adhesion formation or integrin activation studies with activated and inhibitory β3cytoplasmic domain mutants. *J. Biol. Chem.*, **272**(36), 22538–22547.
89. Simone, E., Ding, B.-S., and Muzykantov, V. (2009). Targeted delivery of therapeutics to endothelium. *Cell Tissue Res.*, **335**(1), 283.

90. Steeber, D. A., Venturi, G. M., and Tedder, T. F. (2005). A new twist to the leukocyte adhesion cascade: Intimate cooperation is key. *Trends Immunol.*, **26**(1), 9–12.

91. Bazile, D., Prud'homme, C., Bassoullet, M., Marlard, M., Spenlehauer, G., and Veillard, M. (1995). Stealth Me. PEG-PLA nanoparticles avoid uptake by the mononuclear phagocytes system. *J. Pharm. Sci.*, **84**(4), 493–498.

92. Stein, J., Ries, J., and Barrett, K. E. (1998). Disruption of intestinal barrier function associated with experimental colitis: Possible role of mast cells. *Am. J. Physiol. Liver Physiol.*, **274**(1), G203–G209.

93. Seymour, L. W., Duncan, R., Chytrý, V., Strohalm, J., Ulbrich, K., and Kopeček, J. (1991). Intraperitoneal and subcutaneous retention of a soluble polymeric drug-carrier bearing galactose. *J. Control. Release*, **16**(3), 255–262.

94. Toth, C. A., Thomas, P., Broitman, S. A., and Zamcheck, N. (1985). Receptor-mediated endocytosis of carcinoembryonic antigen by rat liver Kupffer cells. *Cancer Res.*, **45**(1), 392–397.

95. Kaneo, Y., Tanaka, T., Nakano, T., and Yamaguchi, Y. (2001). Evidence for receptor-mediated hepatic uptake of pullulan in rats. *J. Control. Release*, **70**(3), 365–373.

96. Yamaoka, T., Tabata, Y., and Ikada, Y. (1993). Body distribution profile of polysaccharides after intravenous administration. *Drug Deliv.*, **1**(1), 75–82.

97. Davis, G. L., Balart, L. A., Schiff, E. R., Lindsay, K., Bodenheimer Jr, H. C., Perrillo, R. P., Carey, W., Jacobson, I. M., Payne, J., and Dienstag, J. L. (1989). Treatment of chronic hepatitis C with recombinant interferon alfa. *N. Engl. J. Med.*, **321**(22), 1501–1506.

98. Fundueanu, G., Constantin, M., Mihai, D., Bortolotti, F., Cortesi, R., Ascenzi, P., and Menegatti, E. (2003). Pullulan–cyclodextrin microspheres: A chromatographic approach for the evaluation of the drug–cyclodextrin interactions and the determination of the drug release profiles. *J. Chromatogr. B*, **791**(1–2), 407–419.

99. Horiuchi, Y., Hirayama, F., and Uekama, K. (1991). Improvement of stability and bioavailability of 1-hexylcarbamoyl-5-fluorouracil (HCFU) by O-carboxymethyl-O-ethyl-beta-cyclodextrin. *Yakugaku Zasshi J. Pharm. Soc. Japan*, **111**(10), 592–599.

100. Irie, T. and Uekama, K. (1997). Pharmaceutical applications of cyclodextrins. III. Toxicological issues and safety evaluation. *J. Pharm. Sci.*, **86**(2), 147–162.

101. Loftsson, T. and Brewster, M. E. (1996). Pharmaceutical applications of cyclodextrins. 1. Drug solubilization and stabilization. *J. Pharm. Sci.*, **85**(10), 1017–1025.

102. Crini, G., Bertini, S., Torri, G., Naggi, A., Sforzini, D., Vecchi, C., Janus, L., Lekchiri, Y., and Morcellet, M. (1998). Sorption of aromatic compounds in water using insoluble cyclodextrin polymers. *J. Appl. Polym. Sci.*, **68**(12), 1973–1978.

103. Mizobuchi, Y., Tanaka, M., and Shono, T. (1980). Preparation and sorption behaviour of cyclodextrin polyurethane resins. *J. Chromatogr. A*, **194**(2), 153–161.

104. Renard, E., Deratani, A., Volet, G., and Sebille, B. (1997). Preparation and characterization of water soluble high molecular weight β-cyclodextrin-epichlorohydrin polymers. *Eur. Polym. J.*, **33**(1), 49–57.

105. Fundueanu, G., Constantin, M., Oanea, I., Harabagiu, V., Ascenzi, P., and Simionescu, B. C. (2010). Entrapment and release of drugs by a strict "on-off" mechanism in pullulan microspheres with pendant thermosensitive groups. *Biomaterials*, **31**(36), 9544–9553.

106. Chu, L.-Y., Park, S.-H., Yamaguchi, T., and Nakao, S. (2001). Preparation of thermo-responsive core-shell microcapsules with a porous membrane and poly (N-isopropylacrylamide) gates. *J. Memb. Sci.*, **192**(1–2), 27–39.

107. Ding, Z., Fong, R. B., Long, C. J., Stayton, P. S., and Hoffman, A. S. (2001). Size-dependent control of the binding of biotinylated proteins to streptavidin using a polymer shield. *Nature*, **411**(6833), 59.

108. Fundueanu, G., Constantin, M., and Ascenzi, P. (2008). Preparation and characterization of pH-and temperature-sensitive pullulan microspheres for controlled release of drugs. *Biomaterials*, **29**(18), 2767–2775.

109. Maeda, H., Wu, J., Sawa, T., Matsumura, Y., and Hori, K. (2000). Tumor vascular permeability and the EPR effect in macromolecular therapeutics: A review. *J. Control. release*, **65**(1–2), 271–284.

110. Ulbrich, K. and Šubr, V. (2004). Polymeric anticancer drugs with pH-controlled activation. *Adv. Drug Deliv. Rev.*, **56**(7), 1023–1050.

111. Guhagarkar, S. A., Majee, S. B., Samad, A., and Devarajan, P. V. (2011). Evaluation of pullulan-functionalized doxorubicin nanoparticles for asialoglycoprotein receptor-mediated uptake in Hep G2 cell line. *Cancer Nanotechnol.*, **2**(1), 49.

112. Wu, J., Nantz, M. H., and Zern, M. A. (2002). Targeting hepatocytes for drug and gene delivery: Emerging novel approaches and applications. *Front Biosci.*, **7**(3), d717–d725.

113. Nishikawa, T., Akiyoshi, K., and Sunamoto, J. (1994). Supramolecular assembly between nanoparticles of hydrophobized polysaccharide and soluble protein complexation between the self-aggregate of cholesterol-bearing pullulan and alpha-chymotrypsin. *Macromolecules*, **27**(26), 7654–7659.

114. Nishikawa, T., Akiyoshi, K., and Sunamoto, J. (1996). Macromolecular complexation between bovine serum albumin and the self-assembled hydrogel nanoparticle of hydrophobized polysaccharides. *J. Am. Chem. Soc.*, **118**(26), 6110–6115.

115. Akiyoshi, K. and Sunamoto, J. (1996). Supramolecular assembly of hydrophobized polysaccharides. *Supramol. Sci.*, **3**(1–3), 157–163.

116. Akiyoshi, K., Deguchi, S., Moriguchi, N., Yamaguchi, S., and Sunamoto, J. (1993). Self-aggregates of hydrophobized polysaccharides in water. Formation and characteristics of nanoparticles. *Macromolecules*, **26**(12), 3062–3068.

117. Akiyoshi, K., Deguchi, S., Tajima, H., Nishikawa, T., and Sunamoto, J. (1997). Microscopic structure and thermoresponsiveness of a hydrogel nanoparticle by self-assembly of a hydrophobized polysaccharide. *Macromolecules*, **30**(4), 857–861.

118. Deguchi, S., Kuroda, K., Akiyoshi, K., Lindman, B., and Sunamoto, J. (1999). Gelation of cholesterol-bearing pullulan by surfactant and its rheology. *Colloids Surfaces A Physicochem. Eng. Asp.*, **147**(1–2), 203–211.

119. Akiyoshi, K., Kobayashi, S., Shichibe, S., Mix, D., Baudys, M., Kim, S. W., and Sunamoto, J. (1998). Self-assembled hydrogel nanoparticle of cholesterol-bearing pullulan as a carrier of protein drugs: Complexation and stabilization of insulin. *J. Control. Release*, **54**(3), 313–320.

120. Tuncer Degim, I. and Çelebi, N. (2007). Controlled delivery of peptides and proteins. *Curr. Pharm. Des.*, **13**(1), 99–117.

121. Shi, Y. and Li, L. (2005). Current advances in sustained-release systems for parenteral drug delivery. *Expert Opin. Drug Deliv.*, **2**(6), 1039–1058.

122. Boridy, S., Takahashi, H., Akiyoshi, K., and Maysinger, D. (2009). The binding of pullulan modified cholesteryl nanogels to Aβ oligomers and their suppression of cytotoxicity. *Biomaterials*, **30**(29), 5583–5591.

123. Haass, C. and Selkoe, D. J. (2007). Soluble protein oligomers in neurodegeneration: Lessons from the Alzheimer's amyloid β-peptide. *Nat. Rev. Mol. Cell Biol.*, **8**(2), 101.

124. Vinogradov, S. V., Batrakova, E. V., and Kabanov, A. V. (2004). Nanogels for oligonucleotide delivery to the brain. *Bioconjug. Chem.*, **15**(1), 50–60.

125. Satoh, K., Chen, F., Aoyama, A., Date, H., and Akiyoshi, K. (2008). Nanoparticle of cholesterol-bearing pullulan as a carrier of anticancer drugs. *Ejc Suppl.*, **9**(6), 139.

126. Taniguchi, I., Akiyoshi, K., Sunamoto, J., Suda, Y., and Yamamoto, M. (1999). Cell specificity of macromolecular assembly of cholesteryl and galactoside groups-conjugated pullulan. *J. Bioact. Compat. Polym.*, **14**(3), 195–212.

127. Horie, K., Sakagami, M., Kuramochi, K., Hanasaki, K., Hamana, H., and Ito, T. (1999). Enhanced accumulation of sialyl Lewis X-carboxymethylpullulan conjugate in acute inflammatory lesion. *Pharm. Res.*, **16**(2), 314–320.

128. Li, H., Bian, S., Huang, Y., Liang, J., Fan, Y., and Zhang, X. (2014). High drug loading pH-sensitive pullulan-DOX conjugate nanoparticles for hepatic targeting. *J. Biomed. Mater. Res. Part A An Off. J. Soc. Biomater. Japanese Soc. Biomater. Aust. Soc. Biomater. Korean Soc. Biomater.*, **102**(1), 150–159.

129. Lu, D., Wen, X., Liang, J., Gu, Z., Zhang, X., and Fan, Y. (2009). A pH-sensitive nano drug delivery system derived from pullulan/doxorubicin conjugate. *J. Biomed. Mater. Res. Part B Appl. Biomater. An Off. J. Soc. Biomater. Japanese Soc. Biomater. Aust. Soc. Biomater. Korean Soc. Biomater.*, **89**(1), 177–183.

130. Lu, D., Liang, J., Fan, Y., Gu, Z., and Zhang, X. (2010). In vivo evaluation of a pH-sensitive pullulan–doxorubicin conjugate. *Adv. Eng. Mater.*, **12**(9), B496–B503.

131. Masuda, K., Sakagami, M., Horie, K., Nogusa, H., Hamana, H., and Hirano, K. (2001). Evaluation of carboxymethylpullulan as a novel carrier for targeting immune tissues. *Pharm. Res.*, **18**(2), 217–223.

132. Mocanu, G., Mihai, D., Dulong, V., Picton, L., and Lecerf, D. (2011). New anionic amphiphilic thermosensitive pullulan derivatives. *Carbohydr. Polym.*, **84**(1), 276–281.

133. Mocanu, G., Nichifor, M., Picton, L., About-Jaudet, E., and Le Cerf, D. (2014). Preparation and characterization of anionic pullulan thermoassociative nanoparticles for drug delivery. *Carbohydr. Polym.*, **111**, 892–900.

134. Yokoyama, M., Kwon, G. S., Okano, T., Sakurai, Y., and Kataoka, K. (1994). *Development of Micelle-Forming Polymeric Drug with Superior Anticancer Activity*, ACS Publications.

135. Na, K., Lee, K. H., and Bae, Y.H. (2004). pH-sensitivity and pH-dependent interior structural change of self-assembled hydrogel nanoparticles of pullulan acetate/oligo-sulfonamide conjugate. *J. Control. Release*, **97**(3), 513–525.

136. Jung, S.-W., Jeong, Y.-I., and Kim, S.-H. (2003). Characterization of hydrophobized pullulan with various hydrophobicities. *Int. J. Pharm.*, **254**(2), 109–121.

137. Jung, S.-W., Jeong, Y., Kim, Y.-H., and Kim, S.-H. (2004). Self-assembled polymeric nanoparticles of poly (ethylene glycol) grafted pullulan acetate as a novel drug carrier. *Arch. Pharm. Res.*, **27**(5), 562.

138. Park, K.-H., Song, H.-C., Na, K., Bom, H.-S., Lee, K. H., Kim, S., Kang, D., and Lee, D. H. (2007). Ionic strength-sensitive pullulan acetate nanoparticles (PAN) for intratumoral administration of radioisotope: Ionic strength-dependent aggregation behavior and ^{99m}Technetium retention property. *Colloids Surfaces B Biointerfaces*, **59**(1), 16–23.

139. Zhang, H., Li, X., Gao, F., Liu, L., Zhou, Z., and Zhang, Q. (2010). Preparation of folate-modified pullulan acetate nanoparticles for tumor-targeted drug delivery. *Drug Deliv.*, **17**(1), 48–57.

140. Constantin, M., Mihalcea, I., Oanea, I., Harabagiu, V., and Fundueanu, G. (2011). Studies on graft copolymerization of 3-acrylamidopropyl trimethylammonium chloride on pullulan. *Carbohydr. Polym.*, **84**(3), 926–932.

141. Suginoshita, Y., Tabata, Y., Matsumura, T., Toda, Y., Nabeshima, M., Moriyasu, F., Ikada, Y., and Chiba, T. (2002). Liver targeting of human interferon-β with pullulan based on metal coordination. *J. Control. Release*, **83**(1), 75–88.

142. Suginoshita, Y., Tabata, Y., Moriyasu, F., Ikada, Y., and Chiba, T. (2001). Liver targeting of interferon-β with a liver-affinity polysaccharide based on metal coordination in mice. *J. Pharmacol. Exp. Ther.*, **298**(2), 805–811.

143. Cho, J., Park, W., and Na, K. (2009). Self-organized nanogels from pullulan-g-poly (L-lactide) synthesized by one-pot method: Physicochemical characterization and in vitro doxorubicin release. *J. Appl. Polym. Sci.*, **113**(4), 2209–2216.

144. Seo, S., Lee, C.-S., Jung, Y.-S., and Na, K. (2012). Thermo-sensitivity and triggered drug release of polysaccharide nanogels derived from pullulan-g-poly(L-lactide) copolymers. *Carbohydr. Polym.*, **87**(2), 1105–1111.

145. Na, K., Park, K. M., Jo, E. A., and Lee, K. S. (2006). Self-organized pullulan/deoxycholic acid nanogels: Physicochemical characterization

and anti-cancer drug-releasing behavior. *Biotechnol. Bioprocess Eng.*, **11**(3), 262.

146. Lu, Y. and Low, P. S. (2012). Folate-mediated delivery of macromolecular anticancer therapeutic agents. *Adv. Drug Deliv. Rev.*, **64**, 342–352.

147. Li, F., Zhang, H., Gu, C., Fan, L., Qiao, Y., Tao, Y., Cheng, C., Wu, H., and Yi, J. (2013). Self-assembled nanoparticles from folate-decorated maleilated pullulan–doxorubicin conjugate for improved drug delivery to cancer cells. *Polym. Int.*, **62**(2), 165–171.

148. Wang, B., Chen, K., Yang, R., Yang, F., and Liu, J. (2014). Photoresponsive nanogels synthesized using spiropyrane-modified pullulan as potential drug carriers. *J. Appl. Polym. Sci.*, **131**(10).

149. Agasti, S. S., Chompoosor, A., You, C.-C., Ghosh, P., Kim, C. K., and Rotello, V. M. (2009). Photoregulated release of caged anticancer drugs from gold nanoparticles. *J. Am. Chem. Soc.*, **131**(16), 5728–5729.

150. Uda, R. M., Hiraishi, E., Ohnishi, R., Nakahara, Y., and Kimura, K. (2010). Morphological changes in vesicles and release of an encapsulated compound triggered by a photoresponsive malachite green leuconitrile derivative. *Langmuir*, **26**(8), 5444–5450.

151. Zhang, C., An, T., Wang, D., Wan, G., Zhang, M., Wang, H., Zhang, S., Li, R., Yang, X., and Wang, Y. (2016). Stepwise pH-responsive nanoparticles containing charge-reversible pullulan-based shells and poly (β-amino ester)/poly (lactic-co-glycolic acid) cores as carriers of anticancer drugs for combination therapy on hepatocellular carcinoma. *J. Control. Release*, **226**, 193–204.

152. Blechacz, B. and Mishra, L. (2013). Hepatocellular carcinoma biology. In: *Multidisciplinary Treatment of Hepatocellular Carcinoma*, Springer, pp. 1–20.

153. Grenha, A. and Rodrigues, S. (2013). Pullulan-based nanoparticles: Future therapeutic applications in transmucosal protein delivery. *Ther. Deliv.*, **4**(11), 1339–1341.

154. Pawankar, R., Takizawa, R., Goto, M., Goto, Y., Okuda, M., Yamagishi, S., Ohkubo, K., Nonaka, M., Ohtsuka, H., and Yagi, T. (2001). Effect of modified immunotherapy with an allergen–pullulan conjugate in patients with Japanese cedar pollinosis. *Allergol. Int.*, **50**(1), 43–55.

155. Hosseinkhani, H., Aoyama, T., Ogawa, O., and Tabata, Y. (2002). Liver targeting of plasmid DNA by pullulan conjugation based on metal coordination. *J. Control. Release*, **83**(2), 287–302.

156. Thomsen, L. B., Lichota, J., Kim, K. S., and Moos, T. (2011). Gene delivery by pullulan derivatives in brain capillary endothelial cells for protein secretion. *J. Control. Release*, **151**(1), 45–50.

157. Kanatani, I., Ikai, T., Okazaki, A., Jo, J., Yamamoto, M., Imamura, M., Kanematsu, A., Yamamoto, S., Ito, N., and Ogawa, O. (2006). Efficient gene transfer by pullulan–spermine occurs through both clathrin-and raft/caveolae-dependent mechanisms. *J. Control. Release*, **116**(1), 75–82.

158. Nagane, K., Kitada, M., Wakao, S., Dezawa, M., and Tabata, Y. (2009). Practical induction system for dopamine-producing cells from bone marrow stromal cells using spermine-pullulan-mediated reverse transfection method. *Tissue Eng. Part A*, **15**(7), 1655–1665.

159. San Juan, A., Hlawaty, H., Chaubet, F., Letourneur, D., and Feldman, L. J. (2007). Cationized pullulan 3D matrices as new materials for gene transfer. *J. Biomed. Mater. Res. Part A An Off. J. Soc. Biomater. Japanese Soc. Biomater. Aust. Soc. Biomater. Korean Soc. Biomater.*, **82**(2), 354–362.

160. Morimoto, N., Endo, T., Iwasaki, Y., and Akiyoshi, K. (2005). Design of hybrid hydrogels with self-assembled nanogels as cross-linkers: Interaction with proteins and chaperone-like activity. *Biomacromolecules*, **6**(4), 1829–1834.

161. Morimoto, N., Endo, T., Ohtomi, M., Iwasaki, Y., and Akiyoshi, K. (2005). Hybrid nanogels with physical and chemical cross-linking structures as nanocarriers. *Macromol. Biosci.*, **5**(8), 710–716.

162. Akiyoshi, K., Sasaki, Y., and Sunamoto, J. (1999). Molecular chaperone-like activity of hydrogel nanoparticles of hydrophobized pullulan: Thermal stabilization with refolding of carbonic anhydrase B. *Bioconjug. Chem.*, **10**(3), 321–324.

163. Zhang, N., Tan, C., Cai, P., Zhang, P., Zhao, Y., and Jiang, Y. (2009). RNA interference in mammalian cells by siRNAs modified with morpholino nucleoside analogues. *Bioorg. Med. Chem.*, **17**(6), 2441–2446.

164. Kang, J.-H., Tachibana, Y., Kamata, W., Mahara, A., Harada-Shiba, M., and Yamaoka, T. (2010). Liver-targeted siRNA delivery by polyethylenimine (PEI)-pullulan carrier. *Bioorg. Med. Chem.*, **18**(11), 3946–3950.

165. Rekha, M. R. and Sharma, C. P. (2011). Hemocompatible pullulan–polyethyleneimine conjugates for liver cell gene delivery: In vitro evaluation of cellular uptake, intracellular trafficking and transfection efficiency. *Acta Biomater.*, **7**(1), 370–379.

166. Rekha, M. R. and Sharma, C. P. (2009). Blood compatibility and in vitro transfection studies on cationically modified pullulan for liver cell targeted gene delivery. *Biomaterials*, **30**(34), 6655–6664.

167. Na, K., Shin, D., Yun, K., Park, K.-H., and Lee, K. C. (2003). Conjugation of heparin into carboxylated pullulan derivatives as an extracellular matrix for endothelial cell culture. *Biotechnol. Lett.*, **25**(5), 381–385.

168. Tabata, Y. (2003). Tissue regeneration based on drug delivery technology. *Top. Tissue Eng. Finl. Univ. Oulu*, 1–32.

169. Singh, R. S., Kaur, N., Rana, V., and Kennedy, J. F. (2016). Recent insights on applications of pullulan in tissue engineering. *Carbohydr. Polym.*, **153**, 455–462.

170. Perumal, S., Antipova, O., and Orgel, J. P. R. O. (2008). Collagen fibril architecture, domain organization, and triple-helical conformation govern its proteolysis. *Proc. Natl. Acad. Sci.*, **105**(8), 2824–2829.

171. Mallick, K. K. and Cox, S. C. (2013). Biomaterial scaffolds for tissue engineering. *Front Biosci.*, **5**(341), e60.

172. Kumar, D., Saini, N., Pandit, V., and Ali, S. (2012). An insight to pullulan: A biopolymer in pharmaceutical approaches. *Int. J. Basic Appl. Sci.*, **1**(3), 202–219.

173. Chaouat, M., Le Visage, C., Autissier, A., Chaubet, F., and Letourneur, D. (2006). The evaluation of a small-diameter polysaccharide-based arterial graft in rats. *Biomaterials*, **27**(32), 5546–5553.

174. Lack, S., Dulong, V., Le Cerf, D., Picton, L., Argillier, J. F., and Muller, G. (2004). Hydrogels based on pullulan crosslinked with sodium trimetaphosphate (STMP): Rheological study. *Polym. Bull.*, **52**(6), 429–436.

175. Anseth, K. S., Bowman, C. N., and Brannon-Peppas, L. (1996). Mechanical properties of hydrogels and their experimental determination. *Biomaterials*, **17**(17), 1647–1657.

176. Fricain, J. C., Schlaubitz, S., Le Visage, C., Arnault, I., Derkaoui, S. M., Siadous, R., Catros, S., Lalande, C., Bareille, R., and Renard, M. (2013). A nano-hydroxyapatite–pullulan/dextran polysaccharide composite macroporous material for bone tissue engineering. *Biomaterials*, **34**(12), 2947–2959.

177. Bonzi, G., Salmaso, S., Scomparin, A., Eldar-Boock, A., Satchi-Fainaro, R., and Caliceti, P. (2015). Novel pullulan bioconjugate for selective breast cancer bone metastases treatment. *Bioconjug. Chem.*, **26**(3), 489–501.

178. Chen, F., Yu, S., Liu, B., Ni, Y., Yu, C., Su, Y., Zhu, X., Yu, X., Zhou, Y., and Yan, D. (2016). An injectable enzymatically crosslinked carboxymethylated pullulan/chondroitin sulfate hydrogel for cartilage tissue engineering. *Sci. Rep.*, **6**, 20014.

179. Fujioka-Kobayashi, M., Ota, M. S., Shimoda, A., Nakahama, K., Akiyoshi, K., Miyamoto, Y., and Iseki, S. (2012). Cholesteryl group-and acryloyl group-bearing pullulan nanogel to deliver BMP2 and FGF18 for bone tissue engineering. *Biomaterials*, **33**(30), 7613–7620.

180. Barnett, S. E. and Irving, S. J. (1991). Studies of wound healing and the effect of dressings. *High Perform. Biomater.*, 583–620.

181. Lay-Flurrie, K. (2004). The properties of hydrogel dressings and their impact on wound healing. *Prof Nurse*, **19**(5), 269–273.

182. Loke, W., Lau, S., Yong, L. L., Khor, E., and Sum, C. K. (2000). Wound dressing with sustained anti-microbial capability. *J. Biomed. Mater. Res. An Off. J. Soc. Biomater. Japanese Soc. Biomater. Aust. Soc. Biomater. Korean Soc. Biomater.*, **53**(1), 8–17.

183. Li, H., Yang, J., Hu, X., Liang, J., Fan, Y., and Zhang, X. (2011). Superabsorbent polysaccharide hydrogels based on pullulan derivate as antibacterial release wound dressing. *J. Biomed. Mater. Res. Part A*, **98**(1), 31–39.

184. Galvez, M. G., Wong, V. W., Chang, E. I., Major, M., Carre, L., Kandimalla, R., Bhatt, K. A., Rajadas, J., Longaker, M. T., and Gurtner, G. C. (2009). Pullulan-collagen hydrogel scaffold as a dermal substitute. *J. Am. Coll. Surg.*, **209**(3), S78.

185. Wong, V. W., Rustad, K. C., Galvez, M. G., Neofytou, E., Glotzbach, J. P., Januszyk, M., Major, M. R., Sorkin, M., Longaker, M. T., and Rajadas, J. (2010). Engineered pullulan–collagen composite dermal hydrogels improve early cutaneous wound healing. *Tissue Eng. Part A*, **17**(5–6), 631–644.

186. Garg, R. K., Auerbach, L. J., Sorkin, M., Rennert, R. C., Longaker, M. T., and Gurtner, G. C. (2012). A biomimetic collagen-pullulan hydrogel enhances stemness and wound healing potential of adipose-derived mesenchymal stem cells. *J. Am. Coll. Surg.*, **215**(3), S96.

187. Yoshida, Y., Takahata, T., Nakamura, M., Irie, M., Tanaka, M., Ozaki, T., Suzuki, K., and Van Meerbeek, B. (2010). Phosphorylated pullulan bioadhesive for hard-tissue reconstruction. *IADR, Date 2010/07/14-2010/07/17, Locat. Barcelona*.

188. Takahata, T., Okihara, T., Yoshida, Y., Yoshihara, K., Shiozaki, Y., Yoshida, A., Yamane, K., Watanabe, N., Yoshimura, M., and Nakamura, M. (2015). Bone engineering by phosphorylated-pullulan and β-TCP composite. *Biomed. Mater.*, **10**(6), 65009.

189. Li, X., Xue, W., Zhu, C., Fan, D., and Liu, Y. (2015). Novel hydrogels based on carboxyl pullulan and collagen crosslinking with 1,4-butanediol

diglycidylether for use as a dermal filler: Initial in vitro and in vivo investigations. *Mater. Sci. Eng. C*, **57**, 189–196.

190. Atila, D., Keskin, D., and Tezcaner, A. (2015). Cellulose acetate based 3-dimensional electrospun scaffolds for skin tissue engineering applications. *Carbohydr. Polym.*, **133**, 251–261.

191. Kobayashi, H., Katakura, O., Morimoto, N., Akiyoshi, K., and Kasugai, S. (2009). Effects of cholesterol-bearing pullulan (CHP)-nanogels in combination with prostaglandin E1 on wound healing. *J. Biomed. Mater. Res. Part B Appl. Biomater.*, **91**(1), 55–60.

192. Hashimoto, Y., Mukai, S., Sawada, S., Sasaki, Y., and Akiyoshi, K. (2015). Nanogel tectonic porous gel loading biologics, nanocarriers, and cells for advanced scaffold. *Biomaterials*, **37**, 107–115.

193. Schlaubitz, S., Derkaoui, S. M., Marosa, L., Miraux, S., Renard, M., Catros, S., Le Visage, C., Letourneur, D., Amedee, J., and Fricain, J.-C. (2014). Pullulan/dextran/nHA macroporous composite beads for bone repair in a femoral condyle defect in rats. *PLoS One*, **9**(10), e110251.

194. Aydogdu, H., Keskin, D., Baran, E. T., and Tezcaner, A. (2016). Pullulan microcarriers for bone tissue regeneration. *Mater. Sci. Eng. C*, **63**, 439–449.

195. Halstead, F. D., Rauf, M., Bamford, A., Wearn, C. M., Bishop, J. R. B., Burt, R., Fraise, A. P., Moiemen, N. S., Oppenheim, B. A., and Webber, M. A. (2015). Antimicrobial dressings: Comparison of the ability of a panel of dressings to prevent biofilm formation by key burn wound pathogens. *Burns*, **41**(8), 1683–1694.

196. Boonkaew, B., Kempf, M., Kimble, R., Supaphol, P., and Cuttle, L. (2014). Antimicrobial efficacy of a novel silver hydrogel dressing compared to two common silver burn wound dressings: Acticoat™ and PolyMem Silver®. *Burns*, **40**(1), 89–96.

197. Park, M. V. D. Z., Neigh, A. M., Vermeulen, J. P., de la Fonteyne, L. J. J., Verharen, H. W., Briedé, J. J., van Loveren, H., and de Jong, W. H. (2011). The effect of particle size on the cytotoxicity, inflammation, developmental toxicity and genotoxicity of silver nanoparticles. *Biomaterials*, **32**(36), 9810–9817.

198. AshaRani, P. V, Low Kah Mun, G., Hande, M. P., and Valiyaveettil, S. (2008). Cytotoxicity and genotoxicity of silver nanoparticles in human cells. *ACS Nano*, **3**(2), 279–290.

199. Ninan, N., Forget, A., Shastri, V. P., Voelcker, N. H., and Blencowe, A. (2016). Antibacterial and anti-inflammatory pH-responsive tannic acid-carboxylated agarose composite hydrogels for wound healing. *ACS Appl. Mater. Interfaces*, **8**(42), 28511–28521.

200. Xu, F., Weng, B., Gilkerson, R., Materon, L. A., and Lozano, K. (2015). Development of tannic acid/chitosan/pullulan composite nanofibers from aqueous solution for potential applications as wound dressing. *Carbohydr. Polym.*, **115**, 16–24.

201. Román, J. T., Fuenmayor, C. A., Dominguez, C. M. Z., Clavijo-Grimaldo, D., Acosta, M., García-Castañeda, J. E., Fierro-Medina, R., and Rivera-Monroy, Z. J. (2019). Pullulan nanofibers containing the antimicrobial palindromic peptide LfcinB (21–25) Pal obtained via electrospinning. *RSC Adv.*, **9**(35), 20432–20438.

202. Aduba, D. and Yang, H. (2017). Polysaccharide fabrication platforms and biocompatibility assessment as candidate wound dressing materials. *Bioengineering*, **4**(1), 1.

203. Grigoras, A. G. (2019). Drug delivery systems using pullulan, a biocompatible polysaccharide produced by fungal fermentation of starch. *Environ. Chem. Lett.*, 1–15.

204. Choi, Y., Kim, S. Y., Kim, S. H., Lee, K.-S., Kim, C., and Byun, Y. (2001). Long-term delivery of all-trans-retinoic acid using biodegradable PLLA/PEG-PLLA blended microspheres. *Int. J. Pharm.*, **215**(1–2), 67–81.

205. Nogusa, H., Yamamoto, K., Yano, T., Kajiki, M., Hamana, H., and Okuno, S. (2000). Distribution characteristics of carboxymethylpullulan-peptide-doxorubicin conjugates in tumor-bearing rats: Different sequence of peptide spacers and doxorubicin contents. *Biol. Pharm. Bull.*, **23**(5), 621–626.

206. Lee, J., Jeong, D., Seo, S., and Na, K. (2013). Biodegradable nanogel based on all-trans retinoic acid/pullulan conjugate for anti-cancer drug delivery. *J. Pharm. Investig.*, **43**(1), 63–69.

207. Lotan, R. (1995). Retinoids and apoptosis: Implications for cancer chemoprevention and therapy. *J. Natl. Cancer Inst.*, **87**(22), 1655–1657.

208. Zhang, H., Gao, F., Liu, L., Li, X., Zhou, Z., and Zhang, Q. (2009). Pullulan acetate nanoparticles prepared by solvent diffusion method for epirubicin chemotherapy. *Colloids Surfaces B Biointerfaces*, **71**(1), 19–26.

209. Alivisatos, A. P. (1996). Semiconductor clusters, nanocrystals, and quantum dots. *Science*, **271**(5251), 933–937.

210. Arya, H., Kaul, Z., Wadhwa, R., Taira, K., Hirano, T., and Kaul, S. C. (2005). Quantum dots in bio-imaging: Revolution by the small. *Biochem. Biophys. Res. Commun.*, **329**(4), 1173–1177.

211. Chan, W. C. W., Maxwell, D. J., Gao, X., Bailey, R. E., Han, M., and Nie, S. (2002). Luminescent quantum dots for multiplexed biological detection and imaging. *Curr. Opin. Biotechnol.*, **13**(1), 40–46.

212. Michalet, X., Pinaud, F. F., Bentolila, L. A., Tsay, J. M., Doose, S., Li, J. J., Sundaresan, G., Wu, A. M., Gambhir, S. S., and Weiss, S. (2005). Quantum dots for live cells, in vivo imaging, and diagnostics. *Science*, **307**(5709), 538–544.

213. Mitchell, P. (2001). Turning the spotlight on cellular imaging. *Nat. Biotechnol.*, **19**(11), 1013.

214. Derfus, A. M., Chan, W. C. W., and Bhatia, S. N. (2004). Intracellular delivery of quantum dots for live cell labeling and organelle tracking. *Adv. Mater.*, **16**(12), 961–966.

215. Voura, E. B., Jaiswal, J. K., Mattoussi, H., and Simon, S. M. (2004). Tracking metastatic tumor cell extravasation with quantum dot nanocrystals and fluorescence emission-scanning microscopy. *Nat. Med.*, **10**(9), 993.

216. Hasegawa, U., Shin-ichiro, M. N., Kaul, S. C., Hirano, T., and Akiyoshi, K. (2005). Nanogel-quantum dot hybrid nanoparticles for live cell imaging. *Biochem. Biophys. Res. Commun.*, **331**(4), 917–921.

217. Kong, S.-H., Noh, Y.-W., Suh, Y.-S., Park, H. S., Lee, H.-J., Kang, K. W., Kim, H. C., Lim, Y. T., and Yang, H.-K. (2015). Evaluation of the novel near-infrared fluorescence tracers pullulan polymer nanogel and indocyanine green/γ-glutamic acid complex for sentinel lymph node navigation surgery in large animal models. *Gastric Cancer*, **18**(1), 55–64.

218. Yim, H., Yang, S.-G., Jeon, Y. S., Park, I. S., Kim, M., Lee, D. H., Bae, Y. H., and Na, K. (2011). The performance of gadolinium diethylene triamine pentaacetate-pullulan hepatocyte-specific T1 contrast agent for MRI. *Biomaterials*, **32**(22), 5187–5194.

219. Breierová, E., Vajczikova, I., Sasinková, V., Stratilová, E., Fišera, M., Gregor, T., and Šajbidor, J. (2002). Biosorption of cadmium ions by different yeast species. *Zeitschrift für Naturforsch. C*, **57**(7–8), 634–639.

220. Radulović, M. Đ., Cvetković, O. G., Nikolić, S. D., Đorđević, D. S., Jakovljević, D. M., and Vrvić, M. M. (2008). Simultaneous production of pullulan and biosorption of metals by *Aureobasidium pullulans* strain CH-1 on peat hydrolysate. *Bioresour. Technol.*, **99**(14), 6673–6677.

221. Motozato, Y., Ihara, H., Tomoda, T., and Hirayama, C. (1986). Preparation and gel permeation chromatographic properties of pullulan spheres. *J. Chromatogr. A*, **355**, 434–437.

Chapter 6

Functional Characteristics of Pullulan: Its Physicochemical and Physiological Properties

Sanjoy Kumar Das and Soumalya Chakraborty
Institute of Pharmacy, Government of West Bengal,
Jalpaiguri 735101, West Bengal, India
sanjoydasju@gmail.com

Pullulan is a naturally occurring homopolysaccharide of glucose, consisting of repeating subunits of maltotriose, linearly arranged with α-(1→6) glycosidic linkage. It is produced from starch by cultivating the fungus (black yeast) *Aureobasidium pullulans* (*Pullularia pullulans*). It is generally obtained as a white, tasteless, odorless and water-soluble powder with several important applications in pharmaceutical, biomedical, and food industries due to its distinctive physicochemical properties, attributed to the unique nature of chemical linkage present in its structure. Owing to its adhesive nature, it is chiefly used as a film former for oral, oxygen-impermeable biofilm as well as in the formation of fibers and compression molding. Pullulan is widely exploited in the field

Pullulan: Processing, Properties, and Applications
Edited by Shakeel Ahmed and Aisverya Soundararajan
Copyright © 2021 Jenny Stanford Publishing Pte. Ltd.
ISBN 978-981-4877-17-6 (Hardcover), 978-1-003-05694-2 (eBook)
www.jennystanford.com

of tissue engineering and grafting, carrier for drug delivery, medical imaging, plasma volume expander, and as a molecular chaperone. This chapter summarizes the basic physical, chemical, and biochemical properties of pullulan. Additionally, several biological properties and their mechanism in the living world have also been discussed. Furthermore, this chapter discusses properties of the biofilm made from pullulan and the composite film containing pullulan along with other polymers. The different physicochemical characterizations of the pullulan biofilm are also outlined in this chapter.

6.1 Introduction

In the current era biomass, a renewable class of materials is gaining interest among researchers and usually represents a different class of biopolymers [1]. Biopolymers are generally various kinds of plant-based, animal-based, or microbial, water-soluble gums produced from different microorganisms, possessing novel and unique physical properties. Among various biopolymers, different protein-based or polysaccharide-based or sometimes lipid-based biopolymers or their composite polymers are widely used in pharmaceutical, cosmetics, and food industries in the form of emulsifying agent, stabilizing agent, binders, gelling agents, lubricants, film-forming agents, viscosity enhancers, and suspending agents [2]. Pullulan is a linear homopolysaccharide of glucopyranose subunits that can be alternatively called α-(1$\rightarrow$6)-linked maltotriose, primarily a secretion of strains of the fungus *A. pullulans* (often called black yeast due to its property of forming a black pigment, melanin) [3]. Bernier, in the year 1958, first mentioned the existence of pullulan. The chemical structure of pullulan was first elucidated by Bender et al. in 1959. In the year 1976, the Hayashibara Company, located at Okayama, Japan, first began the commercial production of pullulan. Pullulan production was an outgrowth of starch syrup production, noted in 1883. In the year 1982, the Hayashibara Company started commercialization of pullulan films [2]. Pullulan is water soluble with different additional properties, mentioned later, which make it one of the highly suitable materials for edible films and coatings. It is also an ideal material for the formation of water-soluble and biodegradable packaging material [4]. Pullulan films are clear, highly oxygen-impermeable, nontoxic, edible, biodegradable, biocompatible to human and environment, and present good mechanical properties, allowing their use in several fields such as in food and pharmaceutical industries, as coating or packaging materials for dried foods and as coating or component for tablets, pills, and granules, respectively [5]. These oral films are most acceptable band accurate dosage form that could bypass the hepatic system band show more therapeutic response [6]. Owing to their hydrophilic nature, when pullulan films are exposed to elevated relative humidity (RH) condition, substantial reduction of the film occurs. Therefore, the commercially available pullulan-made edible films are highly sensitive to variable humidity conditions, exhibiting brittle nature on exposure to dry environments and becoming sticky under humid conditions. Furthermore, the relatively high cost of pullulan itself compared to other biopolymers such as starch and cellulose derivatives limits its wide utilization for edible films [7]. The blending of polymers is an efficient process for the modification and alteration of the physicochemical characteristics of polymeric materials, making them different from constituent polymers. Actually, polymeric blends are nothing but a physical mixture of the polymers, structurally different from each other, forming a homo-polymeric mixture, by means of secondary force-based interactions like hydrogen bonding, dipole forces, and charge-transfer complex formation. An important property of a polymer blend is the miscibility of its components because it affects the mechanical properties, morphology, and permeability and degradation [8]. Functionalities and properties of pullulan films could be easily tailored by compounding or blending with other polymeric materials such as starch [7, 9], chitosan [10], HPMC [4], caseinate [11], and gellan [12] or by modification of pullulan, for example, with methacrylate [13].

6.2 Physical and Chemical Properties of Pullulan

Dry pullulan is a white to off-white, tasteless, odorless powder forming a viscous non-hygroscopic solution on dissolving in water at 5–10%. The decomposition of pullulan begins at 25°C, and it starts to char at 280°C. It is highly soluble in water and dilute alkali and insoluble in alcohol and other organic solvents except for dimethyl sulfoxide and formamide. Due to its high water solubility, pullulan can be utilized as an efficient carrier material for drugs and also aids in the controlled release of drugs [2]. Aqueous solutions of pullulan are viscous but do not form gels. The viscosity of pullulan solutions is similar to that of the gum arabic and is lower than other soluble polysaccharides, such as guar gum. An aqueous solution of pullulan PI-20 (10% w/w) has a pH of 5.0–7.0 [14]. The specific optical rotation([α]D25) of pullulan is found to be +183 [15]. Infrared spectroscopy of the polysaccharide showed strong absorption at 850 cm^{-l}, characteristic of α-D-glucopyranose units. Absorption at 755 cm^{-1} and from 915 cm^{-1} to 930 cm^{-1} indicated that the predominant linkages were (l→4) and (l→6) glycosidic linkages. The presence of a (1→3) linkage was indicated by a small amount of absorption at 790 cm^{-1} [16]. Derivatization of pullulan has been done to achieve specific purposes. The production of the hydrogenated derivative of pullulan is conducted by the reaction of an aqueous solution of pullulan with hydrogen gas under pressure in the presence of catalysts until the disappearance of the reducing property of the solution. The solubility of pullulan in cold water can be enhanced by carboxylation. Carboxylated pullulan is produced by the oxidation of pullulan until some of the hydroxymethyl groups of the glucose units of pullulan are converted to carboxyl groups. Crosslinking creates an ionic gel able to absorb and retain water to a high degree [17]. Most pullulan derivatizations are intended to reduce its water solubility or to modify its functional behavior by introducing charged or reactive groups. The solubility of pullulan in water can be drastically reduced by esterification or etherification. The heat stability of pullulan can be enhanced by means of hydrogenation [3].

6.3 Biological and Physiological Properties of Pullulan

Pullulan is a slowly digestible polysaccharide in the human body. It was found that the enzymes ptyalin and porcine pancreatic amylase hydrolyzed only 15% and 10% of pullulan during a 22 h *in vitro* incubation. Wolf et al. [18] evaluated the glycemic and breath hydrogen responses and gastrointestinal tolerance to pullulan in comparison to maltodextrin of equivalent amount on diabetic healthy adult subjects and found that the postprandial incremental blood glucose response of nondiabetic healthy adults was improved when subjects consumed pullulan compared with maltodextrin. The extent of *in vitro* maltodextrin hydrolysis was high and rapid; the extent of *in vitro* pullulan hydrolysis was also high, but it occurred at a slow rate in comparison to maltodextrin. Pullulan also increased breath hydrogen excretion and the incidence of gastrointestinal intolerance symptoms, indicating that a portion of pullulan was malabsorbed [18]. Zhang et al. [19] assessed the effect of pullulan on dendritic cell (DC) activation and anticancer immunity in tumor-bearing mice and found that pullulan enhanced DC maturation and function by up-regulating co-stimulatory molecule expression and enhancing pro-inflammatory cytokine production in bone-marrow-derived DCs (BMDC) *in vitro* and in spleen DCs *in vivo* in mice. Pullulan also played the role of an adjuvant in promoting Ag-specific immune responses in mice. The authors concluded that pullulan could be a new and promising adjuvant for use in therapeutic cancer vaccines [19]. Andrews et al. [20] studied the adhesion property of dimorphic fungus *A. pullulans* on liquid and solid media (with varying carbon and nitrogen content) and on the leaf surface. The authors concluded that pullulan is important in the adhesion of *A. pullulans* to leaf surface [20].

6.4 Biodegradable Polymers for Pullulan-Based Composite Film Preparation

Pullulan alone is capable of forming films with good tensile strength, percentage elongation, disintegration time, and dissolution time. The composite films composed of other polymers with pullulan show the opportunity for better film formation with more modified properties. Usually, different proteins are blended with pullulan to prepare biofilms. The biofilm prepared with the protein alone suffers from different limitations, principally poor barrier properties. Several studies showed that the approach of combining protein with pullulan can overcome this limitation by the interaction of these two ingredients with good film-forming characteristics. The resultant films possess extensive barrier properties [21]. Kolaee et al. [22] prepared pullulan–whey protein isolate-based bio-nanocomposite film and evaluated its functional characteristics. It was observed from the works of previous researchers that a combination of polysaccharides and protein may lead to the formation of a film with poor mechanical and barrier properties. It was expected that a bio-nanocomposite approach may overcome this limitation. Initially, whey protein isolate, after dissolving in distilled water and heating at 90°C for 30 min, was added in equal volume solution of pullulan in distilled water at 1:1 w/w. In this solution, 30% w/w glycerol was added as a plasticizer. Montmorillonite was added in dispersion form as a reinforcement agent in the aqueous dispersion of pullulan–whey protein isolate in 1, 3, and 5% w/w concentration, followed by stirring for 1 h. Then the films were prepared by pouring the sample solution on flat and non-adhesive Teflon plates. Appropriate RH and room temperature were maintained. After preparation, the films were separated from the Teflon plates and kept at desiccators [22]. Due to its availability, biodegradability, renewability, film-forming ability, and low cost, starch from different botanical sources (cassava, corn, wheat, rice, potato, pea, and others) is one of the most promising natural polymers for packaging applications. Starch-based films are transparent, odorless, tasteless, and highly impermeable to oxygen. However, they have some limitations such as their highly hydrophilic nature (water sensitivity), which negatively affects tensile and barrier properties as compared with conventional synthetic polymers. Therefore, various approaches, such as blending with additives and/or other biopolymers or starch chemical modification, have been studied to improve these characteristics. Sapper et al. [9] combined pullulan with starch to modify the functional characteristics of starch-made biofilm. It was observed that the pullulan–starch blend polymer showed increased water vapor and oxygen permeability along with no significant alteration of other parameters such as tensile strength and isothermal water sorption capacity. This indicated that the films made up of pullulan–starch blend are inappropriate for serving as packaging material [9]. Sometimes to improve the limitations of poor mechanical barrier properties and transparency of pullulan, a composite film is made by the combination of pullulan with chitosan. Pullulan alone is not suitable for the preparation of edible films as in solution, pullulan exhibits Newtonian flow behavior and is little affected by pH. The physical properties of edible films are highly dependent not only on its chemical composition but pH and solvent as well. Blending pullulan with chitosan imparts the desired pH and solvent sensitivity in the composite film by means of the polar amine groups present in chitosan. In spite of having a large molecular weight (600,000 Da), the significantly low viscosity of pullulan can be attributed to its compact random coil-like conformation. It was observed from the work of Li et al. [10] on the edible films prepared by compositing pullulan and chitosan that the viscosity of the blended film is highly dependent on pH. The authors observed that when pH reduces from 5.0 to 4.0, the viscosity of the blended film increases, probably due to protonation and subsequent repulsion of the amine groups of chitosan, which causes the distortion and unfolding of its compact conformation to form a linear structure. Surprisingly when pH was further reduced to 3.0, the viscosity also reduced slightly, maybe due to the shielding of protonated amines by the chloride ions from HCl solution, used for acidification. The authors further observed that the tensile strength of the film is also inversely proportional to the pH. At reduced pH, protonated chitosan molecule offers more opportunity for hydrogen bond formation leading to enhancement of tensile strength of the film [10]. Usually, the films made by blending pullulan with polysaccharides or proteins show good mechanical properties but are rather sensitive to moisture, due to the hydrophilic nature of these components. On the other hand, films made from lipids show good water vapor barrier properties but are opaque, only slightly flexible, and brittle. This limitation can be overcome by combining the composite films of proteins or polysaccharides with fatty emulsions or fatty layers since the lipids help to lessen water vapor transmission and the proteins or polysaccharides give the necessary film strength and structural integrity. Kristo et al. [11] developed beeswax laminated caseinate–pullulan composite film, which exhibited a drastic reduction in the degree of moisture permeation through the film. Different polymers reported to be used in the preparation of composite film containing pullulan are mentioned along with their chemical nature in Table 6.1.

Table 6.1 Different polymers used for the preparation of composite film containing pullulan

Polymer for composite film preparation	Chemical nature	References
Cassava starch	Plant-based polysaccharide	[9]
Tapioca starch	Plant-based polysaccharide	[7]
Whey protein isolate	Plant-based protein	[21, 22]
Rice protein concentrate	Plant-based protein	[23]
Soy protein concentrate	Plant-based protein	[24]
Chitosan	Chemically modified animal polysaccharide	[10, 25]
Caseinate	Milk protein	[11]
Hydroxy propyl methyl cellulose (HPMC)	Semi-synthetic derivative of natural polysaccharide	[4]
Nanofibrilated cellulose	Chemically modified natural polysaccharide	[5]
Gellan	Microbial polysaccharide	[12]
Kappa-carrageenan	Marine polysaccharide	[1]
Rice wax	Lipid	[23]
Graphene oxide	Chemically modified graphene, containing oxygen functional groups such as epoxides, alcohols, and carboxylic acids	[26]
Micro-fibrillated cellulose	Chemically modified polysaccharide	[27]
Polymethacrylate	Synthetic biocompatible polymer	[13]

6.5 Physicochemical Characterization of Pullulan Biofilm

Different physicochemical characterizations of pullulan biofilm are discussed in the following paragraphs.

6.5.1 Film Thickness

Usually, film thickness is measured after the permeability study to avoid any damage to the film during film thickness measurement. A micrometer caliper is used to estimate film thickness. Usually, an average of nine readings taken arbitrarily on the film samples gives the value of the overall thickness [21].

6.5.2 Film Color

A colorimeter is usually used to estimate the color of the film. The color of samples is determined according to the lab system against a white background as standard, in which *L*, *a*, and *b* describe white–black, red–green, and yellow–blue, respectively. The total color difference (ΔE) (Eq. 6.1), yellow index (YI) (Eq. 6.2), and white index (WI) (Eq. 6.3) were calculated using the following equations:

$$\Delta E = \sqrt{(L^* - L)^2 + (a^* - a)^2 + (b^* - b)^2} \tag{6.1}$$

$$\mathrm{YI} = \frac{142.86 \times b}{L} \tag{6.2}$$

$$\mathrm{WI} = 100 - \sqrt{(100 - L)^2 + (a)^2 + (b)^2} \tag{6.3}$$

where L^*, a^*, and b^* are the color-parameter values of the standard and *L*, *a*, and *b* are the color-parameter values of the specimen. All measurements are usually repeated five times [22].

6.5.3 Water Vapor Permeability and Water Vapor Transmission Rate

Water vapor permeability and water vapor transmission rate are determined by the gravimetric method. The films, after keeping at 53% RH at 25°C for 48 h, are cut into disk-shaped pieces and loaded into the cups containing distilled water. The cups are sealed and exposed to a condition of 25°C and 90% RH. The condition is maintained with the help of a saturated solution of $MgCl_2 \cdot 6H_2O$. A saturated solution of $BaCl_2$ can also be used for this purpose. At every 2 h interval, weight change is monitored, and it is continued for 24 h. A graph is plotted between weight increase versus temperature. The gradient is equal to the slope of the regression line [11]. From Eq. 6.4, film permeance is determined.

$$\text{Permeance} = \frac{\text{Slope}}{A\,.\,\Delta p} \tag{6.4}$$

where A is the exposed area of the film and Δp is the water vapor pressure differential across the film [11].

Equation 6.5 [11] provides the value of water vapor permeability (WVP):

$$\text{WVP} = \text{Permeance} \times \text{Thickness} \tag{6.5}$$

6.5.4 Moisture Content

To determine the moisture content, square-shaped specimens of the films are taken and kept in a desiccator containing a saturated solution of potassium acetate, magnesium nitrate, and potassium chloride for a constant RH of 23, 53, and 85%, respectively. Temperature is maintained at 25°C. The samples are usually stored for 6 days. Then the moisture content of the film is estimated by drying at 105°C in an oven for 3 h [7].

6.5.5 Water Solubility

The water solubility of the film is defined as the percentage of dry matter present in the film, which becomes soluble after keeping the film immersed in distilled water for a period of 6 h. To determine the water solubility of the film, initially the film is cut into pieces with definite dimension and put into an oven at 103±2°C for a period of 3 h, to evaporate the moisture content previously present in it and make the weight of the film constant. Then the dried films are immersed in distilled water with periodic agitation for 6 h at 25°C. After the specified period, the remaining undissolved pieces of the film are collected and dried at the previous condition and time duration to make it constant weight. The same procedure is generally repeated thrice, and the moisture content is evaluated from Eq. 6.6 [24].

$$\text{WS} = \frac{m_2 - m_1}{m_1} \times 100 \tag{6.6}$$

where m_1 is the weight of the film before dissolving in distilled water, and m_2 is the weight of the film after dissolving in distilled water.

Very often different crosslinking agents (e.g., glutaraldehyde) are included in the formulation of the film to increase its tensile strength. To estimate the degree of crosslinking, usually the swelling of the film after immersing in distilled water is monitored. Generally, it is considered that the higher the degree of swelling, the lower the degree of crosslinking [1].

6.5.6 Mechanical Properties

Mechanical properties of pullulan films are important to estimate as it might be exposed to different kinds of stress during handling and use. The mechanical properties of pullulan biofilm include tensile strength and elongation at break. These properties are evaluated by using a Testometric machine. The desiccator containing saturated magnesium nitrate solution was used for conditioning film specimens in RH of 55% at 25±1°C for at least 48 h [24]. At least three rectangular ribbons are cut and fixed at a grip distance of 50 mm. A head speed of 10 mm/min is fixed for the evaluation of film samples [22]. Tensile strength and elongation at break are calculated using Eqs. 6.7 and 6.8, respectively:

$$\mathrm{TS} = \frac{F_{\mathrm{Max}}}{A_{\mathrm{Min}}} \tag{6.7}$$

$$\mathrm{EB} = \frac{L_{\mathrm{Max}}}{L_0} \times 100 \tag{6.8}$$

where F_{max} is the maximum load, A_{min} is the minimum area of cross section, L_{max} is the extension at the moment of rupture, and L_0 is the initial length of the specimen [22].

6.5.7 Surface Morphology

Field emission scanning electron microscopy is usually used to evaluate both surface morphology and the texture of the cross-sectional area of the film [24].

6.5.8 Transparency

Usually, a UV–Vis spectrophotometer operating at the wavelength 200–800 nm is employed to estimate the barrier property of the film against UV light and visible light. Equation 6.9 provides a measure of the transparency of the film:

$$\text{Transparency} = {}^{A_{600}}\!/\!_{X} \tag{6.9}$$

where A_{600} is absorbance at 600 nm wavelength and X is the thickness of the film in mm [21].

6.5.9 Mass Transfer Flask

The mass transfer characteristic of a membrane is a function of the concentration of the solution passed through a membrane at a given time. To properly understand the complex process of mass transfer, it is important to estimate the diffusion coefficient of the membrane, which is given by Eq. 6.10:

$$D_{\mathrm{W}} = \frac{1.86 \times 10^{-18} (\varphi_{\mathrm{B}} M_{\mathrm{W}})^{0.5}}{\mu_{\mathrm{W}} v_{\mathrm{A}^{0.6}}} T \tag{6.10}$$

where D_{W} is the diffusion coefficient of the solute in water [m^2/s], μ_{W} is the viscosity of water [Pa.s], and [K] is the association factor for solvent B at the required temperature T (for water, $\varphi_{\mathrm{B}} = 2.6$). M_{W} is the molar mass of water [g/mol], and V_{A} is the molar volume of solute A at the normal boiling point [m^3/mol] [1].

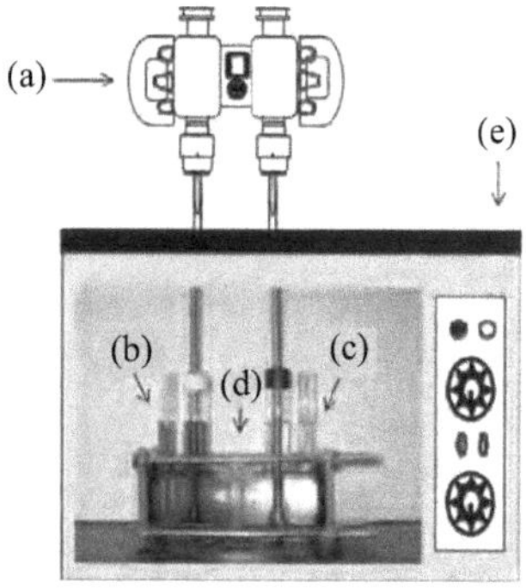

Figure 6.1 Schematic diagram of experimental setup of mass transfer coefficient determination by Wu and Imai (2011). Here, (a) agitating motor, (b) mass-transfer cell (feed side), (c) mass-transfer cell (stripping side), (e) membrane, (d) constant temperature water tank (303 K) (adapted from Ref. [1]).

Wu and Imai [1], in their experimental setup, fixed the membrane in between two equivolume chambers filled with the same volume of distilled water (Fig. 6.1). In one of the compartments, a water-soluble diffusion marker was added. Both compartments are stirred at a constant rate to minimize the resistance toward mass transfer offered by soluble boundary layer at the membrane. An aliquot of the sample was drawn from both compartments at a regular interval and was analyzed by using a UV–Vis spectrophotometer. From the collected data, the diffusion coefficient was estimated with a slight modification of the previously mentioned equation [1].

6.5.10 Surface Topography

The surface topography of the film is studied by using an atomic force microscope (DualScope or Rasterscope). For analysis, prior to scanning, the sample film is cut into appropriate pieces and fitted onto the sample stage by using double-sided adhesive tape. After scanning, all necessary image analysis and parametric interpretation are done by employing a DualScope or Rasterscope SPM software [22]. The mean surface roughness (S_a) is estimated by using Eq. 6.11 [22].

$$S_a = \iint_a |Z(x, y)| dxdy \tag{6.11}$$

For calculating the mean square roughness (S_q), Eq. 6.12 [22] is used.

$$S_q = \sqrt{\iint_a (Z(x, y))^2 dxdy} \tag{6.12}$$

6.5.11 Oxygen Permeability

The oxygen permeation rate is determined according to the ASTM standard method D3985-05. The films are previously equilibrated at 25°C and 53% RH at which further measurements were taken. The transmission values are determined every 20 min interval. The process is continued until equilibrium is established. The thickness and exposure area of the film are considered important parameters during the estimation of the oxygen permeation rate [9]. For a more detailed knowledge about this matter, one may look through ASTM, "Standard test method for oxygen gas transmission rate through plastic film and sheeting using a colorimetric sensor. Standard designations: 3985-3995," in the Annual Book of ASTM Standards, 2002.

6.6 Conclusion

In the recent years, pullulan has gained immense attention among researchers owing to its unique structural properties and the possibilities of its use in various sectors of pharmaceutical, food, biomedical, and packaging industries. This carbohydrate-based biopolymer is also known for its excellent film-forming properties, especially in the case of edible biofilms. These films can be prepared with the pullulan alone or by combining with different polymers to prepare composite films. This approach of the blending of polymers intended for structural and functional modification of the native polymer is exploited in the case of the research regarding pullulan films. Pullulan films are not still widely commercialized, mainly due to different drawbacks like high cost in comparison to other used polymers, lack of rigidity in humid exposure due to its hydrophilic nature, storage issues, etc. Most of the research work on pullulan composite film is based on the combination of pullulan with other biopolymers. The field of studies regarding the combination of pullulan with synthetic biomimetic polymers is not still widely exploited. Therefore, more research works should be conducted in these sectors.

Acknowledgments

The authors sincerely thank Dr. Rajan Rajabalaya of Universiti Brunei Darussalam, who provided valuable advice and help for the preparation of the chapter.

References

1. Wu, P. and Imai, M. (2011). Food polymer pullulan-carrageenan composite membrane performed smart function both on mass transfer and molecular size recognition, *Desalin. Water. Treat.*, **34**, pp. 239–245.
2. Prajapati, V. D., Jani, G. K., and Khanda, S. M. (2013). Pullulan: An exopolysaccharide and its various applications, *Carbohydr. Polym.*, **95**, pp. 540–549.
3. Leathers, T. D. (2003). Biotechnological production and application of pullulan, *Appl. Microbiol. Biotechnol.*, **62**, pp. 468–473.

4. Yan, J., Li, Z., Zhang, J., and Qiao, C. (2012). Preparation and properties of pullulan composite films, *Adv. Mater. Res.*, **476**, pp. 2100–2104.

5. Trovatti, E., Fernandes, S. C. M., Rubatat, L., Perez, D. S. P., Freire, C. S. R., Silverstre, A. J. D., and Neto, C. P. (2012). Pullulan–nanofibrillated cellulose composite films with improved thermal and mechanical properties, *Composit. Sci. Technol.*, **72**, pp. 1556–1561.

6. Ganduri, V. S. R. K., Reethika, N. M., Jayasai, M., Sirisha, V., Vasudha, S., and Poda, S. (2016). Effect of pullulan concentration in fast dissolving films formulation and exploration of film properties, *J. Pharm. Res.*, **10**, pp. 211–215.

7. Kim, J., Choi, Y., Kim, S. R. B., and Lim, S. T. (2014). Humidity stability of tapioca starch–pullulan composite films, *Food. Hydrocoll.*, **41**, pp. 140–145.

8. Islam, M. S., Oh, W., Shin, J. C., Lee, J. T., and Yeum, J. H. (2014). Poly(vinyl alcohol)/pullulan blend nanofibres prepared from aqueous solutions using electrospinning method, *Polym. Polym. Compos.*, **22**, pp. 770–779.

9. Sapper, M., Talens, P., and Chiralt, A. (2019). Improving functional properties of cassava starch-based films by incorporating xanthan, gellan, or pullulan gums, *Int. J. Polym. Sci.*, **2019**, pp. 1–8.

10. Li, Y., Yokoyama, W., Wu, J., Ma, J., and Zhong, F. (2015). Properties of edible films based on pullulan–chitosan blended film-forming solutions at different pH, *RSC Adv.*, **2015**, pp. 105844–105850.

11. Kristo, E., Biliaderis, C. G., and Zampraka, A. (2007). Water vapour barrier and tensile properties of composite caseinate–pullulan films: Biopolymer composition effects and impact of beeswax lamination, *Food. Chem.*, **101**, pp. 753–764.

12. Zhu, G., Sheng, L., Li, J., and Tongi, Q. (2013). Preparation and characterization of gellan/pullulan composite blend films, *Int. J. food. Sci. Technol.*, **48**, pp. 2683–2687.

13. Lima, I. A. D. L., Pomin, S. P., and Cavalcanti, O. A. (2017). Development and characterization of pullulan–polymethacrylate free films as potential material for enteric drug release, *Braz. J. Pharm. Sci.*, **e00002**, pp. 1–9.

14. Stankovic, I. (2005). Pullulan, *Proc. Chemical and Technical Assessment 65th JECFA.*

15. Bashir, M. and Kassim, I. (2000). Isolation and characterization of pullulan produced by a local isolate of *Aureobasidium pullulans*, *Qatar. Univ. Sci. J.*, **20**, pp. 105–110.

16. Sow, W., Blackwood, A. C., and Adam, G. A. (1963). Natural extracellular glucan of *Pullularia pullulans* (de Bary) berkhouti, *Canadian. J. Chem.*, **41**, pp. 2314–2319.

17. LeDuy, A., Choplin, L., Zajic, J. E., and Luong, J. H. T. (2014). Pullulan: Properties, synthesis, and applications, *Encyclopedia of Polymer Science and Technology*, https://doi.org/10.1002/0471440264.pst620.

18. Wolf, B. W., Garleb, K. A., Choe, Y. S., Humphrey, P. M., and Maki, K. C. (2003). Pullulan is a slowly digestible carbohydrate in humans, *J. Nutr.*, **133**, pp. 1051–1055.

19. Zhang, W., Yu, X., Kwak, M., Su, L., Zhang, L., and Yu, Q. (2016). Maturation of dendritic cells by pullulan promotes anti-cancer effect, *Oncotarget.*, **7**, pp. 1051–1055.

20. Andrews, J. H., Harris, R. F., Spear, R. N., Lau, G. W., and Nordhiem, E. V. (1994). Morphogenesis and adhesion of *Aureobasidium pullulans*, *Can. J. Microbiol.*, **40**, pp. 6–17.

21. Gounga, M. E., Xu, S., and Wang, Z. (2007). Whey protein isolate-based edible films as affected by protein concentration, glycerol ratio and pullulan addition in film formation, *J. Food. Eng.*, **83**, pp. 521–530.

22. Kolaee, M., Khodaiyan, F., and Ghahfarrokhi, I. (2016). Modification of functional properties of pullulan–whey protein bionanocomposite films with nanoclay, *J. Food. Sci. Technol.*, **53**, pp. 1294–1302.

23. Shih, F. F. (1996). Edible films from rice protein concentrate and pullulan, *Cereal. Chem.*, **73**, pp. 406–409.

24. Rad, F. H., Khodaiyan, F., Hosseini, S. E., and Sharifan, A. (2018). Preparation and characterization of pullulan-soy protein concentrate biocomposite film, *J. Food. Biosci. Technol.*, **8**, pp. 19–28.

25. Lazaridou, A., Biliaderis, C. G., and Kontogiorgos, V. K. (2003). Molecular weight effects on solution rheology of pullulan and mechanical properties of its films, *Carb. Polym.*, **52**, pp. 151–166.

26. Unalan, I. U., Wan, C., Figiel, Ł., Olsson, R. T., Trabattoni, S., and Farris, S. (2015). Exceptional oxygen barrier performance of pullulan nanocomposites with ultra-low loading of graphene oxide, *Nanotechnology*, **26**(27), 275703.

27. Cozzolino, C. A., Cerri, G., Brundu, A., and Farris, S. (2014). Microfibrillated cellulose (MFC): Pullulan bionanocomposite films, *Cellulose*, **21**, pp. 4323–4335.

Chapter 7

Pullulan-Degrading Enzymes and Their Biochemical Features

Srijita Basumallick
Department of Chemistry, Asutosh College under Calcutta University, 92, S P Mukherjee Road, Kolkata 700026, India
srijitabasumallick@gmail.com

Pullulan is a highly ordered branched polysaccharide comprising linear chains of D-glucopyranosyl units, which arrange in between one (1–6) α D link and two (1–4) α D links. Pullulan-degrading enzymes (PDEs), known as pullulanase, break the polymer to smaller oligomers, hydrolyzing glycoside linkages. In this chapter, we discuss different types of pullulanase, their origin (plants/animals/fungi/bacteria), and some common biochemical features with special reference to substrate-binding domains. In addition to these, basic enzymatic features such as substrate specificity and product analysis have also been presented. An attempt has been made to present different mechanisms proposed from time to time on enzymatic degradation of pullulan. Applications of enzymatic hydrolysis of pullulan in the industry for manufacturing maltotriose and glucose have also been discussed.

Pullulan: Processing, Properties, and Applications
Edited by Shakeel Ahmed and Aisverya Soundararajan
Copyright © 2021 Jenny Stanford Publishing Pte. Ltd.
ISBN 978-981-4877-17-6 (Hardcover), 978-1-003-05694-2 (eBook)
www.jennystanford.com

7.1 Introduction

Pullulan is a highly ordered branched polysaccharide [1–5] comprising linear chains of D-glucopyranosyl units arranged in between one (1–6) α D link and two (1–4) α D links (Fig. 7.1). Thus, pullulan-degrading enzymes (PDEs) have the ability [6–8] to hydrolyze these glycosidic linkages. Enzyme pullulanase (pullulan α-glucanohydrolase; enzyme EC.3.2.1.41) is an extracellular carbohydrase, which was first isolated by Bender and Wallenfels [9] in 1966 from the mesophilic organism *Klebsiella pneumoniae.* Apart from this enzyme catalytic (EC.3.2.1.41) identification, there are pullulanases known as isopullulanase (EC.3.2.1.57) [7, 8] and neopullulanase (EC.3.2.1.35) [7, 8]. These pullulanases can be subdivided into two types: pullulanase type-I and pullulanase type-II. Pullulanase type-I hydrolyzes only α-1,6-glycocidic linkage, but pullulanase type-II specifically attacks on α-1,4-glycocidic linkage and α-1,6-glycocidic linkage in some selected carbohydrate.

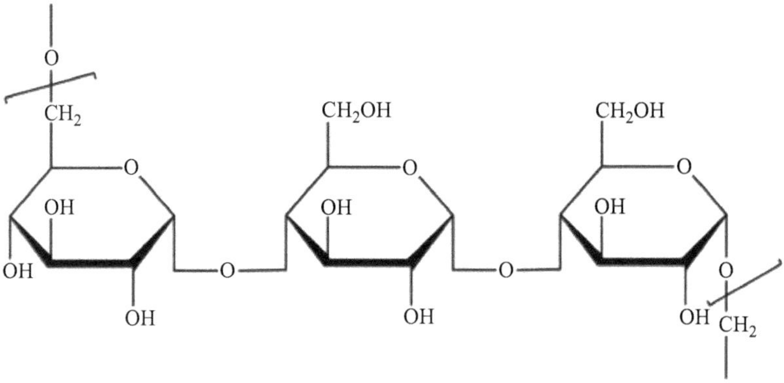

Figure 7.1 Pictorial representation of pullulan.

Pullulanases are also called de-branching enzymes, which hydrolyze the branch chain polysaccharides. PDEs can be further classified into four groups [7] based on substrate specificities and reaction products: (i) pullulan hydrolase type-I, which attacks α-1,4-glycosidic linkages in pullulan, forming panose (it was previously classified as a neopullulanase as stated above); (ii) pullulan hydrolase type-II attacks α-1,4-glycosidic linkages in pullulan, forming isopanose (it was previously classified as isopullulanase); (iii) pullulanase type-I specifically hydrolyzes α-1,6-glycosidic linkages in pullulan or branched oligosaccharides, forming maltotriose or linear oligomers, respectively; and (iv) pullulanase type-II attacks α-1,6-glycosidic linkages in pullulan and branched substrates in addition to the α-1,4-glycosidic linkages in polysaccharides other than pullulan. Apart from these pullulanases, type-III pullulan hydrolase from *Thermococcus kodakarensis* (TK-PUL) possesses both pullulanase and amylase activities. Until now, only two enzymes in this class [10], which are capable of hydrolyzing both α -1,4- and 1,6-glycosidic bonds in pullulan to produce a mixture of maltose, panose, and maltotriose, have been described. A summary of these activities is presented in Table 7.1.

A pictorial representation of enzymatic action on pullulan is presented in Fig. 7.2.

The TK-PUL structure and function of the type-III pullulan hydrolase from *Thermococcus kodakarensis* [10] show the highest activity in the temperature range 95–100°C and have a pH optimum in the range 3.5–4.2. Its ability to hydrolyze maltotriose into maltose and glucose has not been found for other homologous enzymes. The crystal structure of TK-PUL [10] has been determined at a resolution of 2.8 Å and represents the first analysis of a type-III pullulan hydrolyze. The structure reveals that the last parts of the N-terminal domain and the C-terminal domain are significantly different from homologous structures. In addition, the loop regions at the active-site end of the central catalytic domain are quite different. The enzyme has a well-defined calcium-binding site and possesses a rare vicinal [11]. In the recent years, a large number of pullulanases have been isolated, particularly from thermophilic microorganisms. The research on pullulanase from thermophiles is interesting not only for understanding thermal stability [12] of the enzyme but also for exploring improved enzymes for application in the industrial starch hydrolysis process. The initial step in the conversion of starch to glucose is liquefaction, which runs at temperatures of 95°C to 105°C in the presence of a thermostable α-amylase. This is followed by saccharification at a temperature of about 60°C using a pullulanase type-I (debranching enzyme) and glucoamylase. The presence of a thermostable debranching enzyme that is active above 80°C [10] will significantly improve the starch bioconversion process. The pullulanases studied in detail to date are type-II enzymes

Table 7.1 Reaction specificities of pullulan-degrading enzymes

Enzyme	EC number	Bonds processed	Preferred substrate	End products	References
Pullulanase type I	3.2.1.41	α-(1,6)	Oligo- and polysaccharides, pullulan	Trimer (maltotriose)	[17, 27]
Pullulanase type II (amylopullulanase)	3.2.1.41	α-(1,6)	Pullulan	Trimer (maltotriose)	[31, 32, 40]
		α-(1,4)	Poly- and oligosaccharide (starch)	Mixture of glucose, maltose, and maltotriose	
Pullulan hydrolase type I (neopullulanase)	3.2.1.135	α-(1,4)	Pullulan	Panose	[41–43]
Pullulan hydrolase type II (isopullulanase)	3.2.1.57	α-(1,4)	Pullulan	Isopanose	[12]
Pullulan hydrolase type III	3.2.1—	α-(1,4) and α-(1,6)	Pullulan	Mixture of panose, maltose, and maltotriose	[37]
			Starch, amylose, and amylopectin	Maltotriose and maltose	

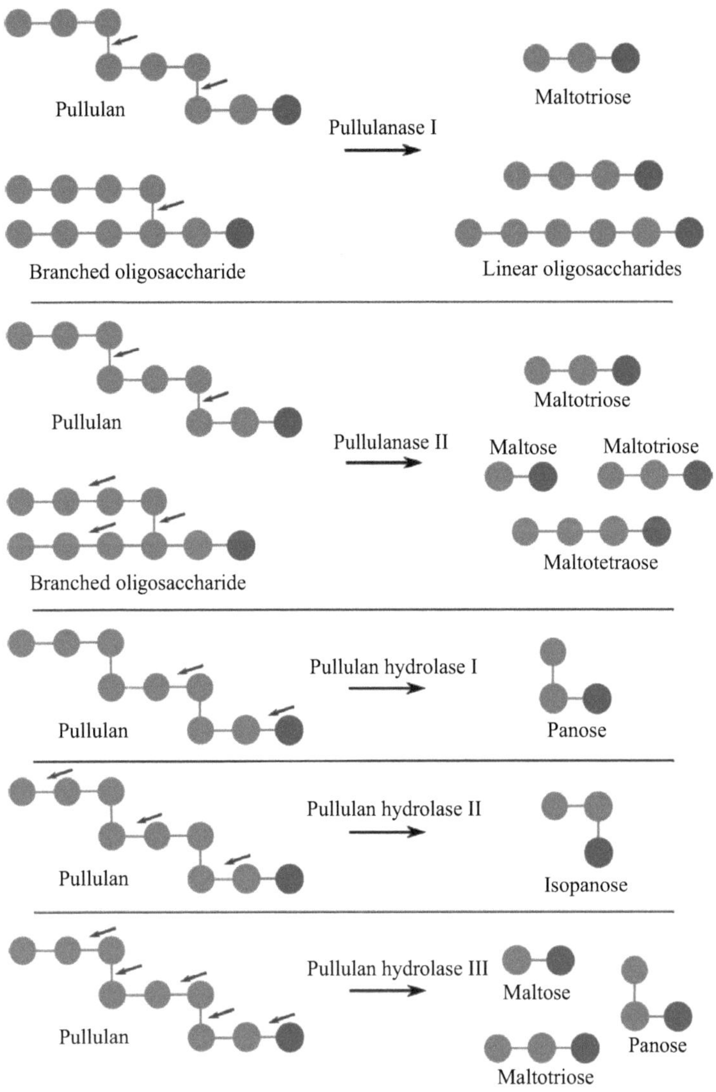

Figure 7.2 A pictorial representation of degradation of pullulan by different pullulanases. Reproduced from Ref. [10] with permission.

and are derived from extreme thermophilic archaea. Pullulanase obtained from plant sources is known as limit dextranase (EC.3.2.1.42). It was previously called R-enzyme. It is known [13] that pullulanase type-I and isoamylase type of enzymes are involved in determination of actual structure of amylopectin. After the isolation of pullulanase type-I (referred here as pullulanase) from a culture of *Aerobacter aerogenes* (*K. pneumoniae*) by Wallenfels et al. in 1966 [9], the enzyme has been isolated from several microorganisms, including *Streptococcus mitis*, Bacillus no. 202-1, and *Klebsiella aerogenes* W70, and characterized [14]. *K. pneumoniae* pullulanase [14] is composed of 1150 amino acid residues and has at least four domains [10] consisting of catalytic and C-terminal domains similar to those of α-amylase. The catalytic domain includes four conserved regions, I to IV, which are common to the enzymes of the α-amylase family, together with the consensus sequence of YNWGYDP found in pullulanase type-I. The mutational study of the role of conserved residues of *K. aerogenes* pullulanase showed that the substitution of His607, Asp677, and His682 led to the complete loss of enzyme activity, and that the substitution of His607 and Asp677 resulted in the loss of the ability to bind cyclodextrins, which are competitive inhibitors of pullulanase. In spite of these studies about the structure and function of pullulanase, many things remain unexplained, owing to the lack of proper understanding of the three-dimensional structure of pullulanase.

7.2 Enzyme Activity of Pullulanase on Other Polysaccharides

Pullulanase type-I and type-II as well as pullulan hydrolase can not only degrade pullulan but can also degrade starch to maltotriose and maltose together with malto-oligosaccharides [15, 16] with an end product of amylose. Similarly, it can hydrolyze (1–6)-α-D-glucosidic linkages in amylopectin producing malto-oligosaccharides, maltose [17], and glycogen, and in the alpha- and beta-limit dextrins of amylopectin and glycogen and maltotriose [18], malto-2-4-oligosaccharide and other product [19]. Pullulanase type-I can also degrade amylose to malto-2-4-oligosaccharide and other products [19]. It can also degrade glycogen to maltose [17, 20], maltotriose [20], and malto-oligosaccharides [17] and malto-2-4-oligosaccharide

[19]. It can also degrade γ-cyclodextrin [21, 22] and β-cyclodextrin [21]. However, pullulanase type-I does not hydrolyze either α-1,4-linkages in glucans and amylose or α-1,6-linkages in glycogen and panose [23]. Both pullulanase type-I and II enzymes cannot degrade cyclodextrin [21].

7.3 Crystal Structure of Pullulanase

The crystal structures of pullulanase (from *K. pneumoniae*) and its complex with glucose, maltose, isomaltose, maltotriose, and maltotetraose have been refined at around 1.7–1.9 Å resolution by using a synchrotron radiation source [24]. The refined models contained 920–1052 amino acid residues, 942–1212 water molecules, four or five calcium ions, and the bound sugar moieties. Pullulanase type-I is composed of five domains (N1, N2, N3, A, and C). Among all these, CBM (carbohydrate-binding domain)-41 or N1 is most important as it helps in substrate binding. Very interestingly this domain is absent [25] or is disordered in crystal [26]. In pullulanase (from *K. pneumoniae*), the N1 domain was clearly visible only in the structure of the complex with G3 or G4. The N1 and N2 domains are characteristic of pullulanase, while the N3, A, and C domains have weak similarity with those of *Pseudomonas isoamylase*. The N1 domain was found to be a new type of carbohydrate-binding domain with one calcium site (CBM41). All five glucose, maltose, isomaltose, maltotriose, and maltotetraose units bind at different subsites (Fig. 7.3b); for example, two maltotetraose bind in −1 ~ −4 and +2 ~ −1′ subsite, indicating anchoring from both sides of the catalytic site of maltotetraose. Very importantly this subsite has close proximity with catalytic residues Glu706 and Asp677, indicating their roll in glycosidase chemistry (Fig. 7.3b). Also these subsites together with catalytic residues Glu706 and Asp677 are conserved between pullulanase and α-amylase, which indicates pullulanase-branched sugar and the point of branching. Very interestingly Mikami et al. found that two maltotriose and maltotetraose molecules are in parallel position and interact with each other in a dimmer of enzyme protein [24]. The comparison suggested that the conformational difference around the active cleft, together with the domain organization, determines the different substrate specificities between pullulanase and isoamylase. Very interestingly, the starch-binding domains in glucoamylase, cyclodextrin glucanotransferase, and β-amylase (CBM20) have different amino acid sequences [27] as well as different tertiary and secondary structure [28] compared to the sugar-binding site of pullulanase (CBM41). Active site of these enzymes contain two Trp, one Lys and one Gln (ASp). Some important interactions in the active site of pullulanase (CBM41) are Arg889 (two hydrogen bond through NH) and Thr642 (hydrogen bond through O) with the catalytic site of the maltotriose. Other interactions with the non-reducing end of maltotriose, making two hydrogen bonds ($O2/O^{\delta 1}$ of Asp138 and $O6/O^{\eta}$ of Tyr78), and the reducing end, making two hydrogen bonds ($O2/N^{\zeta}$ of Lys133 and $O3/N^{\zeta}$ of Lys133) [24], are also important. The active sites of all the different types of pullulanases have similar crystal structure as well as the active site with Glu and Asp residues playing an important role in catalytic activity.

Here it is worth mentioning that both pullulanase type-II (amylopullulanase) and neopullulanase have one single active site to break both 1–4 and 1–6 glycosidic linkages [29]. Very interestingly in neopullulanase, if Glu357, Asp424, or Asp328 were replaced with His or Gln and with His or Asn, the neopullulanase loses its activity [30]. His423, Glu332, and Asn331 are also known to involve in substrate recognition [30]. Recently [30], it has been seen from Kuriki et al. that the main amino acid residues distinguishing neopullulanase from amylopullulanase are Ala357, Gln359, and Tyr360 residues and 11 other consecutive amino acids of region II. By changing these amino acids, the neopullulanase can only hydrolyze the 1–6 bond. These researchers have shown that two Asp and one Glu of neopullulanase participate in the dual activity toward α-(1,4)- and α-(1,6)-glucosidic linkages [30]. The Ile358 of neopullulanase can alter substrate specificity and the trans-glycosylation activity of neopullulanase [31].

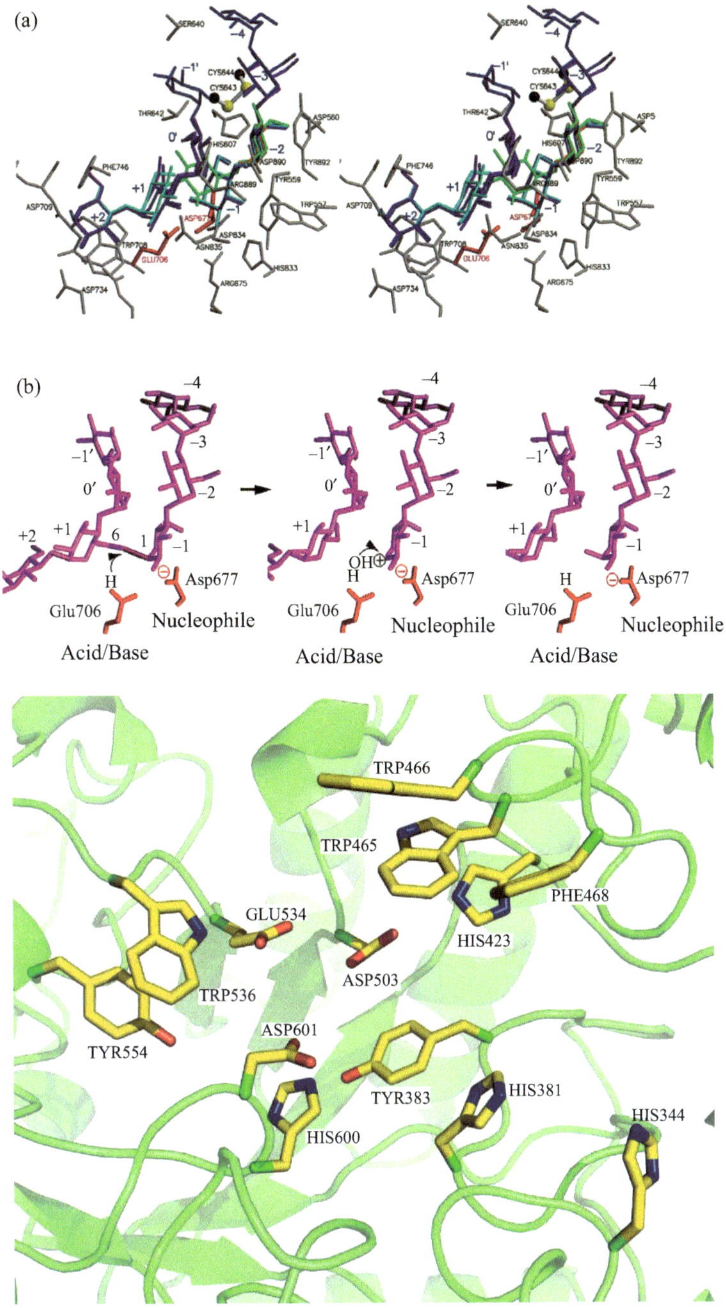

Figure 7.3 The active site of TK-PUL showing the catalytic triad, and aromatic residues that are involved in carbohydrate binding are shown in ball-and-stick representation. Reproduced from Ref. [10] with permission from the author.

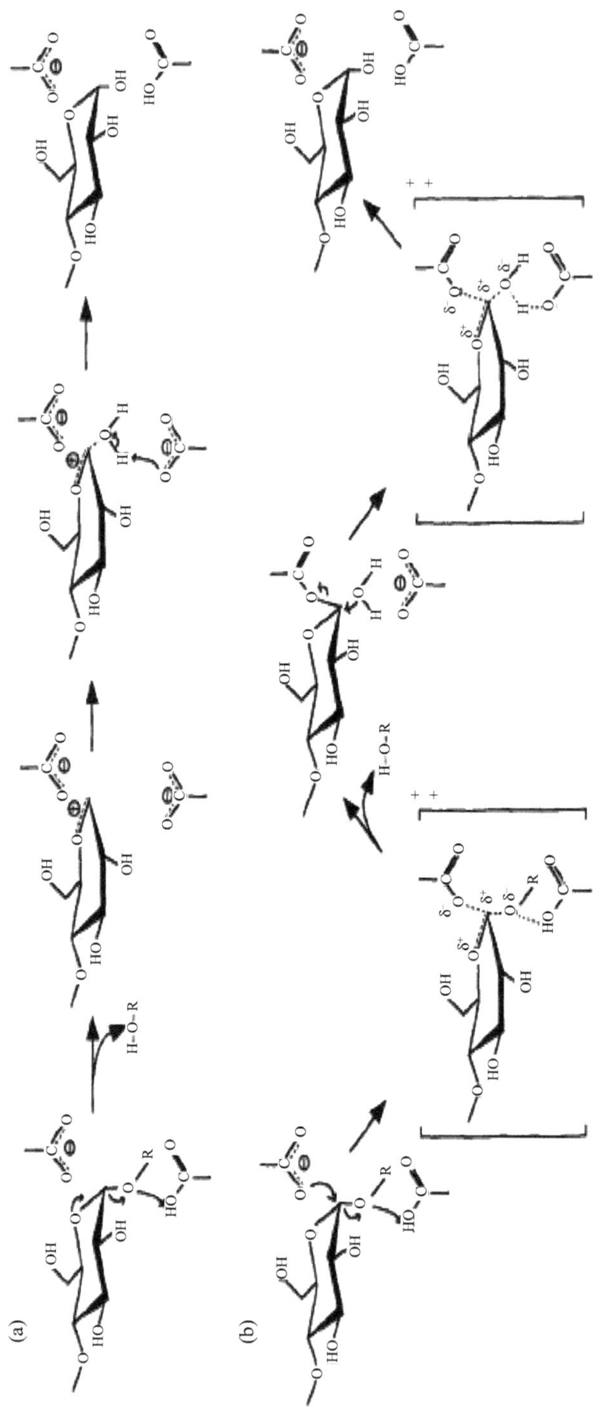

Figure 7.4 (a) Enzymatic degradation by SN1 mechanism (b) by SN2 mechanism.

(a)

(b)

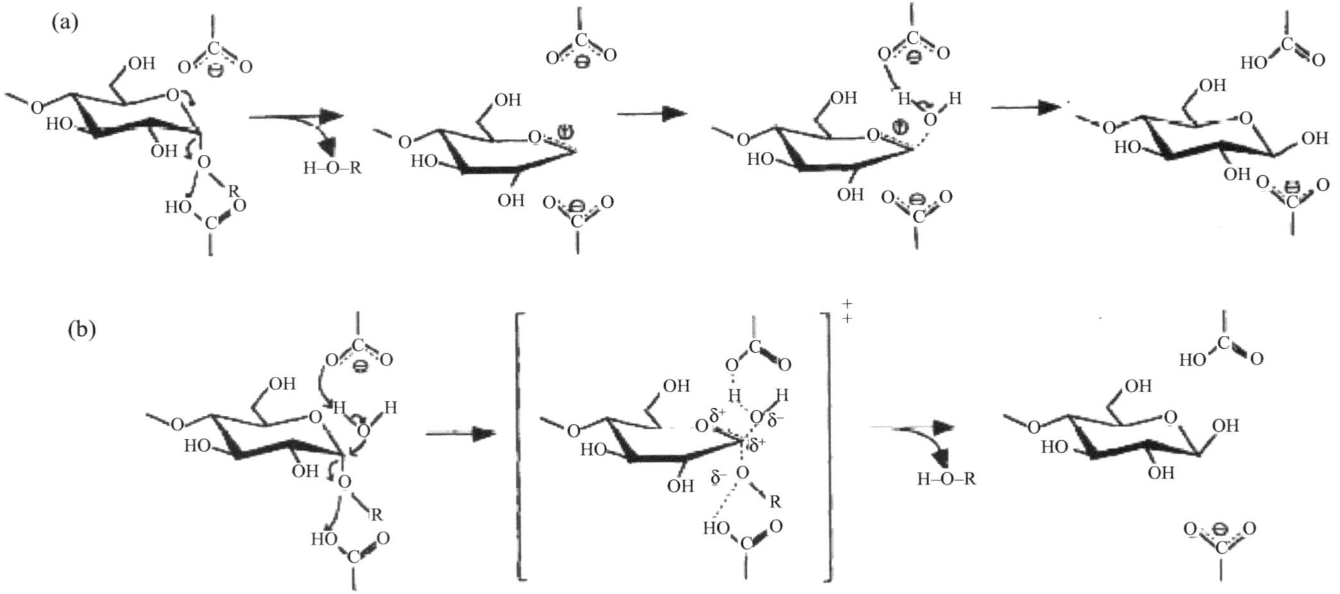

Figure 7.5 Enzymatic degradation of pullulan (a) SN1 product with different configuration, (b) similar configuration for SN2.

7.4 Mechanism of Glycosidic Linkage Hydrolysis

Here it is worth mentioning that all glucose structures found in the active site of the enzyme are in the $^4C_1$-pyranoside forms [24]. There are two mechanisms proposed, one SN1 (Figs. 7.4a and 7.5a) and another SN2 (Figs. 7.4b and 7.5b) forming either alpha or beta anomer of the final product.

In the case of pullulanase enzymes, it is speculated [29] and also supported [10, 24] that SN1 reaction will occur at the actual active site of enzymes. The proton from a Glu residue attacks the glycosidic bond, leading to the formation of a carbocation that undergoes nucleophilic attack by the carboxyl group of Asp residue followed by molecular bond formation. In the next step, water molecule attacks in a different SN2 manner from the other side of the Asp group as shown in Figs. 7.6 and 7.3b.

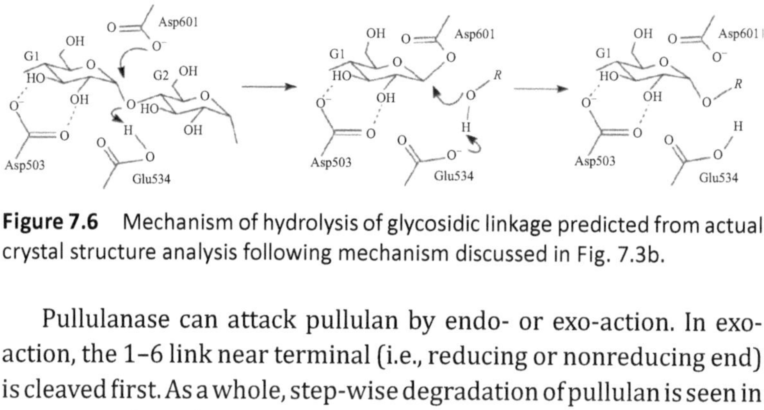

Figure 7.6 Mechanism of hydrolysis of glycosidic linkage predicted from actual crystal structure analysis following mechanism discussed in Fig. 7.3b.

Pullulanase can attack pullulan by endo- or exo-action. In exo-action, the 1–6 link near terminal (i.e., reducing or nonreducing end) is cleaved first. As a whole, step-wise degradation of pullulan is seen in exo-action. In endo-action, any 1–6 link in the middle of the polymer chain can be cleaved, producing two big fragments. Pullulanases type-I typically hydrolyze in an endo-way on the 1,6-α-D-glucosidic linkages in pullulan and in branched oligosaccharides to produce maltotriose and linear oligosaccharides, respectively.

7.5 Enzymatic Properties of Pullulanase

Since pullulanase type-II has been studied extensively, we present here enzymatic properties of this important enzyme. Pullulanase is one of the important enzymes in the starch industry. The search for low-cost pullulanase with distinct features [32], possibly from easily grown bacterium, is of interest for industrial applications. The extracellular pullulanase produced by *Bacillus cereus* HI.5 is generally purified by easy chromatographic method of DEAE-Sepharose, followed by Superdex gel filtration. After purification, the enzyme is characterized in terms of the optimal pH and temperature for activity as well as substrate specificity. It has been reported that the enzyme showed optimal activity at 55°C and pH 6.0. It has also been reported [32] that the thermo-stability and the thermo-activity of the enzyme are increased considerably in the presence of Ca^{2+}. In the presence of 2 mM Ca^{2+}, the enzyme had half-life duration of more than 2 h at 50°C. Almost all metal ions had a strong inhibitory effect, except Ca^{2+} and Mn^{2+}. The Ca^{2+} had a very strong stimulating effect on the enzyme, increasing its activity by 170%. The enzyme is activated by 2-mercaptoethanol and dithiothreitol, whereas N-bromo-succinimide and Schardinger dextrins act as inhibitors, suggesting that tryptophan and thiol residues may be important for the enzyme activity. The apparent K_m and V_{max} values for pullulan were 1.1 mg/mL and 0.275 μmol/min, respectively. The substrate specificities of pullulanase enzyme for hydrolysis of pullulan, amylopectin, and soluble starch were 100%, 28.5%, and 20.4%, respectively. The enzyme was able to attack specifically the α-1,6 linkages in pullulan to generate maltotriose as the major end product, as well as the α-1,4 linkages in amylopectin and soluble starch leading to the formation of a mixture of maltose and glucose and, therefore, be classified as a type-II pullulanase or an amylopullulanase [32].

7.6 Industrial Application of Pullulanase

Earlier starch hydrolysis was carried out by traditional acid hydrolysis method. Since the 1960s, almost all processes to convert starch to glucose have changed to the enzymatic hydrolysis method. Today, much of the starch hydrolysates available in the market are enzyme-converted products of higher dextrose equivalent (DE) value [33]. DE is a measure of reducing sugar on a dry basis, pure dextrose (glucose) being DE 100 and starch being close to DE 0. The enzymatic conversion of starch into glucose, maltose, and fructose for use as food sweeteners is a rapidly growing area in industrial enzyme usage. The primary application of pullulanase is in starch saccharification, and the most important industrial use of pullulanase is in the production of high-glucose (30% to 50% glucose; 30% to 40% maltose) or high-maltose (30% to 50% maltose; 6% to 10% glucose) syrups [34, 35]. In the saccharification process, pullulanase is normally used in combination with glucoamylase or β-amylase [14, 35]. In the hydrolysis of starch, pullulanase and isoamylase play important roles in living cells too because they are the only enzymes that selectively hydrolyze α-1,6 linkages in glucans. Both enzymes are endo acting and produce α-1,4 glucan chains or malto-oligosaccharides. Pullulanase hydrolyzes pullulan and amylopectin completely or nearly completely; however, it can hydrolyze glycogen only to a limited extent. On the other hand, isoamylase hydrolyzes amylopectin and glycogen completely; however, it cannot act on pullulan. Isoamylase is used as a powerful tool in the analysis of the fine structure of amylopectin, whereas pullulanase, in combination with β-amylase, is widely used in the starch saccharification industry to produce maltose. Enzymatic properties of isoamylase from *Pseudomonas* amyloderamosa have been extensively studied, and an enzyme preparation from *Pseudomonas* sp. is commercially available. Pullulanase preparations from *Klebsiella* sp., *Bacillus brevis*, *B. licheniformis*, and *Bacillus acidopulluliticus* are available from several enzyme manufacturers. The enzyme prepared from *B. acidopulluliticus* is thermostable and thus widely used in the starch industry.

7.7 Other Applications

Pullulanase also finds minor application in the manufacturing of low-calorie beer and in the baking industry as an antistaling agent to improve texture, volume, and flavor of bakery products [34]. It is also used as a dental plaque control agent.

7.8 Conclusion

This chapter presents an overview of different types of pullulanase and their capability of hydrolyzing glycosidic linkages of polysaccharides. The active sites of these pullulanases have been analyzed from the analysis of their crystal structures. Mechanisms of hydrolysis of glycosidic linkages have been briefly discussed. These are widely used in saccharification process for the production of various useful food materials such as maltose, amylose, and glucose, by debranching starch with and without α-amylase, β-amylase, or glucoamylase, respectively. Thus, pullulanase can facilitate a major change in the current strategy for starch-processing industries. Apart from its industrial applications, basic understanding on its substrate specificity has also been highlighted.

Acknowledgment

The author is thankful to the Principal and Head of the Department of Chemistry of Asutosh College for proving necessary facilities and encouragement.

References

1. Trinetta, V. and Cutter, C. N. (2016). Pullulan: A suitable biopolymer for antimicrobial food packaging applications. In: *Antimicrobial Food Packaging* (Barros-Velázquez, J., Ed.), Academic Press, San Diego, pp. 385–397.
2. LeDuy, A., Choplin, L., Zajic, J. E., and Luong, J. H. T. (2014). Pullulan: Properties, synthesis, and applications, In: *Encyclopedia of Polymer Science and Technology*, DOI: 10.1002/0471440264.pst620.
3. Oğuzhan, P. and Yangılar, F. (2013). Pullulan: Production and usage in food industry, *Afr. J. Food Sci. Technol.*, **4**(3), pp. 57–633.
4. Bender, H. W. (1961). Investigations on pullulan. II. Specific degradation by means of a bacterial enzyme, *Biochem. Z*, **334**, pp. 79–95.
5. Bernier, B. (1958). The production of polysaccharides by fungi active in the decomposition of wood and forest litter, *Can. J. Microbiol.*, **4**, pp. 195–204.
6. Doman-Pytka, M. and Bardowski, J. (2004). Pullulan degrading enzymes of bacterial origin, *Crit. Rev. Microbiol.*, **30**, pp. 107–121.
7. Hii, S. L., Tan, J. S., Ling, T. C., and Ariff, A. B. (2012). Pullulanase: Role in starch hydrolysis and potential industrial applications, *Enzyme Res.*, http://dx.doi.org/10.1155/2012/921362.

8. Malakar, R., Tiwari, A., and Malviya, S. N. (2010). Pullulanase: A potential enzyme for industrial application, *IJBR*, pp. 10–20, DOI: https://doi.org/10.7439/ijbr.v1i2.53.
9. Bender, H. and Wallenfels, K. (1966). Pullulanase (an amylopectin and glycogen debranching enzyme) from *Aerobacter aerogenes.* In: *Methods in Enzymology*, Academic Press, pp. 555–559.
10. Guo, J., Coker, A. R., Wood, S. P., Cooper, J. P., Keegan, R. M., Ahmad, N., Muhammad, M. A., Rashid, N., and Akhtar, M. (2018). Structure and function of the type III pullulan hydrolase from *Thermococcus kodakarensis, Acta Crystallographica Section D*, **74**, pp. 305–314.
11. Carugo, O., Cemazar, M., Zahariev, S., Hudaky, I., Gaspari, Z., Perczel, A., and Pongor, S. (2003). Vicinal disulfide turns, *Protein Eng.*, **16**, pp. 637–639.
12. Rudiger, A., Jorgensen, P. L., and Antranikian, G. (1995). Isolation and characterization of a heat-stable pullulanase from the hyperthermophilic archaeon *Pyrococcus woesei* after cloning and expression of its gene in *Escherichia coli, Appl. Env. Microbiol.*, **61**, pp. 567–575.
13. Sharma, B., Pickens, J. B., Striegler, S., and Barnett, J. D. (2018). Biomimetic glycoside hydrolysis by a microgel templated with a competitive glycosidase inhibitor, *ACS Catal.*, **8**, pp. 8788–8795.
14. Haki, G. D. and Rakshit, S. K. (2003). Developments in industrially important thermostable enzymes: A review, *Bioresour. Technol.*, **89**, pp. 17–34.
15. Qiao, Y., Peng, Q., Yan, J., Wang, H., Ding, H., and Shi, B. (2015). Gene cloning and enzymatic characterization of alkali-tolerant type I pullulanase from *Exiguobacterium acetylicum*, *Lett. Appl. Microbiol.*, **60**, pp. 52–59.
16. Kang, S., Vieille, C., and Zeikus, J. G. (2004). Identification of *Pyrococcus furiosus* amylopullulanase catalytic residues, *Appl. Microbiol. Biotechnol.*, **66**, pp. 408–413.
17. Yamasaki, Y., Nakashima, S., and Konno, H. (2008). Pullulanase from rice endosperm, *Acta Biochimica Polonica*, **55**, pp. 507–510.
18. Bertoldo, C., Armbrecht, M., Becker, F., Schäfer, T., Antranikian, G., and Liebl, W. (2004). Cloning, sequencing, and characterization of a heat- and alkali-stable type I pullulanase from *Anaerobranca gottschalkii, Appl. Environ. Microbiol.*, **70**, pp. 3407.
19. Lin, F. L. (2002). Cloning, expression, and characterization of thermostable region of amylopullulanase gene from

Thermoanaerobacter ethanolicus 39E, *Appl. Biochem. Biotechnol.*, **97**, pp. 33–44.

20. Lin, F. P., Ma, H. Y., Lin, H. J., Liu, S. M., and Tzou, W. S. (2011). Biochemical characterization of two truncated forms of amylopullulanase from *Thermoanaerobacterium saccharolyticum* NTOU1 to identify its enzymatically active region, *Appl. Biochem. Biotechnol.*, **165**, pp. 1047–1056.
21. Duffner, F., Bertoldo, C., Andersen, J. T., Wagner, K., and Antranikian, G. (2000). A new thermoactive pullulanase from *Desulfurococcus mucosus*: Cloning, sequencing, purification, and characterization of the recombinant enzyme after expression in *Bacillus subtilis*, *J. Bacteriol.*, **182**, pp. 6331-6338.
22. Singh, R. S., Saini, G. K., and Kennedy, J. F. (2010). Maltotriose syrup preparation from pullulan using pullulanase, *Carbohydr. Polym.*, **80**, pp. 401–407.
23. Kashiwabara, S., Ogawa, S., Miyoshi, N., Oda, M., and Suzuki, Y. (1999). Three domains comprised in thermostable molecular weight 54,000 pullulanase of type I from *Bacillus flavocaldarius* KP1228, *Biosci. Biotechnol. Biochem.*, **63**, pp. 1736–1748.
24. Mikami, B., Iwamoto, H., Malle, D., Yoon, H., Demirkan, E., Mezaki, Y., and Katsuya, Y. (2006). Crystal structure of pullulanase: Evidence for parallel binding of oligosaccharides in the active site, *J. Mol. Biol.*, **359**, pp. 690–707.
25. Gourlay, L. J., Santi, I., Pezzicoli, A., Grandi, G., Soriani, M., and Bolognesi, M. (2009). Group B streptococcus pullulanase crystal structures in the context of a novel strategy for vaccine development, *J. Bacteriol.*, **191**, pp. 3544–3552.
26. Turkenburg, J., Brzozowski, A. M., Svendsen, A., Borchert, T. V., Davies, G., and Wilson, K. S. (2006). Structure of a pullulanase from *Bacillus acidopullulyticus*, *Proteins*, **76**, pp. 516–519.
27. Rodriguez-Sanoja, R., Oviedo, N., and Sanchez, S. (2005). Microbial starch-binding domain, *Curr. Opin. Microbiol.*, **8**, pp. 260–267.
28. Mikami, B., Adachi, M., Kage, T., Sarikaya, E., Nanmori, T., Shinke, R., and Utsumi, S. (1999). Structure of raw starch-digesting *Bacillus cereus* β-amylase complexed with maltose, *Biochemistry*, **38**, pp. 7050–7061.
29. Pytka, M. J. and Bardowski, J. (2004). Pullulan degrading enzymes of bacterial origin, *Crit. Rev. Microbiol.*, **30**(2), pp. 107–121.
30. Kuriki, T., Takata, H., Okada, S., and Imanaka, T. (1991). Analysis of the active center of *Bacillus stearothermophilus* neopullulanase, *J. Bacteriol.*, **173**, pp. 6147–6152.

31. Kuriki, T., Kaneko, H., Yanase, M., Takata, H., Shimada, J., Handa, S., Takada, T., Umeyama, H., and Okada, S. (1996). Controlling substrate preference and transglycosylation activity of neopullulanase by manipulating steric constraint and hydrophobicity in active center, *J. Biol. Chem.*, **271**, pp. 17321–17329.

32. Ling, H. S., Ling, T. C., Mohamad, R., and Ariff, A. B. (2009). Characterization of pullulanase type II from *Bacillus cereus* H1.5, *Am. J. Biochem. Biotechnol.*, **5**, pp. 170–179.

33. Zareian, S., Khajeh, K., Ranjbar, B., Dabirmanesh, B., Ghollasi, M., and Mollania, N. (2010). Purification and characterization of a novel amylopullulanase that converts pullulan to glucose, maltose, and maltotriose and starch to glucose and maltose, *Enzyme Microb. Technol.*, **46**, pp. 57–63.

34. Marc., J., Cvan, E., and Dijkhuizen, L. (2002). Properties and applications of starch-converting enzymes of the α-amylase family, *J. Biotechnol.*, **94**, pp. 137–155.

35. Gomes, I., Gomes, J., and Steiner, W. (2003). Highly thermostable amylase and pullulanase of the extreme thermophilic eubacterium *Rhodothermus marinus*: Production and partial characterization, *Bioresour. Technol.*, **90**, pp. 207–214.

Chapter 8

Microbial Pullulan: Properties, Bioprocess Engineering, and Applications

Sugumaran Karuppiah, Sameeha Syed Abdul Rahman, and V. Ponnusami

Bioprocess Engineering Laboratory & Centre for Bioenergy, School of Chemical and Biotechnology, SASTRA Deemed to be University, Thirumalaisamudram 613402

sugumaran@biotech.sastra.edu; chembiotsugu@gmail.com

A variety of drug carriers, namely, liposomes, biodegradable polysaccharides, chemically modified polymer nanoparticles, and microspheres, are employed in drug-delivery applications for improving the degree of therapeutic effect and reducing the toxicity effect. Microbial exopolysaccharides, namely, dextran, xanthan gum, and pullulan, contribute attractive and potential applications due to their biodegradability and biocompatibility. Microbial pullulan has gained attraction of researchers among other polysaccharides due to the following reasons: (1) desirable chemical structure and molecular weight; (2) superior mechanical strength, thermal

Pullulan: Processing, Properties, and Applications
Edited by Shakeel Ahmed and Aisverya Soundararajan
Copyright © 2021 Jenny Stanford Publishing Pte. Ltd.
ISBN 978-981-4877-17-6 (Hardcover), 978-1-003-05694-2 (eBook)
www.jennystanford.com

stability, rheological properties; and (3) increasing applications of modified pullulan in various industrial sectors as a drug/gene carrier, scaffold material, encapsulation matrix, and food packing material. The biosynthesis of pullulan using *Aureobasidium pullulans*, which has the key enzymes, namely, UDPG pyrophosphorylase, α-phosphoglucomutase, and glucosyl transferase, is investigated in various biochemical reactors (submerged fermentation or solid-state fermentation). In this chapter, properties, bioprocess, purification of pullulan, and industrial applications of pullulan derivatives are elaborately discussed.

8.1 Introduction

Proteins, polyesters, and polysaccharides are different classes of polymers that are synthesized by living organisms. Microbes secrete them as metabolites during fermentation, and they have unique physical properties and chemical compositions [1]. Polysaccharides can be broadly classified, on the basis of repeating units, as (1) homopolysaccharides, which have same monomers that are glucans and fructans, and (2) heteropolysaccharides, which have two or more monosaccharides [2]. Depending on the mode of secretion of the metabolites during fermentation by microbes, they can be further categorized as (1) intracellular polysaccharides, which are synthesized within the cells, and (2) exopolysaccharides, which are secreted outside the cells [3, 4]. The produced polysaccharides during the fermentation can be either soluble or precipitated in the production medium, which will affect the downstream processing cost.

Owing to their unique chemical structure, rheological property, and physicochemical properties such as molecular weight and mechanical strength, the microbial exopolysaccharides are potentially employed in various fields in food, pharmaceutical, agriculture, cosmetics, textile industries, and tissue engineering.

The microbial polysaccharides, namely, dextran, bacterial cellulose, pullulan, xanthan gum, gellan gum, welan gum, and levan, are industrially important metabolites, used in various industrial sectors due to their specific physical property and chemical composition [5].

Dextran, an extracellular polysaccharide of D-glucopyranose, is the first approved polysaccharide by the Food and Drug Administration (FDA) [6]. It is a homopolysaccharide produced by *Leuconostoc mesenteroides* and is extensively used in biochemical, pharma, and food industries as a viscosifier, stabilizer, carrier, and emulsifier [7]. Xanthan gum, an extracellular heteropolysaccharide, biosynthesized by *Xanthomonas campesteris*, is the second approved biodegradable polysaccharide by the FDA as a food constituent [8]. It is widely used in the food industry. In that list, pullulan is a safe polysaccharide and received the status of Generally Regarded as Safe (GRAS) in the United States [9]. Nowadays, chemically modified pullulan derivatives from pullulan are extensively applied as a drug vehicle in the pharmaceutical industry [10, 11]. During the formation of thin film from pullulan solution, it seems to be transparent, resistant to oxygen transfer, and used for food packaging and pharmaceutical coating materials in industries [12]. It is used as an alternative for starch in low calorific food and as an internal standard molecular weight marker for evaluating unknown component molecular weight in size exclusion and high-performance liquid chromatography (HPLC) [13]. Pullulan has also been used in various applications such as flocculants, blood plasma substitutes, cosmetic additives, and food additives [14, 15]. In spite of its widespread importance in the food, pharmaceutical, and other fields, its usage at large scale is limited due to technoeconomic constraints.

The production of pullulan is carried out either in batch or continuous cultivation of *Aureobasidium pullulans*. It has the following morphological structure: (1) chlamydospores, (2) yeast-like cells, (3) swollen blastospores, (4) mycelia, and (5) young blastospores [16]. The morphology of cell is highly influenced by initial pH, age of fermentation, and medium composition and, in turn, affects the production of pullulan. Even though the presence of polymorphic characteristics of *A. pullulans* due to various factors such as pH/nitrogen source/carbon source, the yeast-like morphology is suitable for the production of pullulan [16–18]. Generally, the nature of substrate, type of microorganism, mode of fermentation, and purification strategy can influence the chemical structure as well as the molecular weight of pullulan [19]. Because of its unique chemical composition, degree of polymerization, physicochemical characteristics, biocompatibility, and biodegradability, pullulan has increasing applications in various industrial sectors such as food and biomedical engineering. In recent publications, the cost of pullulan has been reported about three times higher than that of other polysaccharides such as levan, xanthan gum, and dextran [1], which is mainly due to production cost. Pullulan costs about ca. $25 per kg, which is relatively high compared to other polysaccharides. This cost might affect the usage in various industrial fields. Around 30% of the total production cost is contributed by the media composition [20]. In order to bring down the cost of pullulan production, numerous approaches are adopted, such as innovations in engineering techniques, improved strains [1], and cost-effective carbon and nitrogen sources [21, 22]. Therefore, the production of microbial pullulan under economical aspect is a challenging task. In this chapter, properties, bioprocess, purification, and industrial applications of microbial pullulan are elaborately discussed.

8.2 Properties

8.2.1 Physical Properties

Pullulan has the following physicochemical properties represented in Table 8.1.

Table 8.1 Physicochemical parameters of pullulan

S. No.	Parameter	Specification	References
1	Appearance	White or off-white	[1]
2	Nature	Powder	[23]
3	pH (1% w/v solution)	5.0–7.0	[1]
4	Decomposition point	250–280°C	[24]
5	Glass transition temperature	82.3°C	[25]
6	Molecular weight	45–600 kDa	[26]
7	Viscosity (10% (w/v) solution, at 30°C)	100–180 cP	[23]

S. No.	Parameter	Specification	References
8	Precipitation capability	Organic solvent, namely, methanol, ethanol, isopropanol, and acetone, except DMF and DMSO	[23]
9	Moisture (loss on drying in weight percentage)	Approximately 6%	[23]
10	Specific optical activity [α] D_2O (1% (w/v) in water)	Less than or equal to +160°	[1]
11	Moisture content (absence of tackiness or hygroscopy)	10–15%	[23]
12	Polypeptides	Less than 0.5%	[23]
13	Mineral residue ash (sulfated)	Less than 3%	[23]

Pullulan is a colorless, odorless, nonionic, nontoxic, nonmutagenic, noncarcinogenic, and nonhygroscopic powder. It is a water-soluble polysaccharide that forms clear and adhesive solution. It is insoluble in organic solvents such as methanol, ethanol, and isopropanol, except dimethylformamide (DMF) and dimethyl sulfoxide (DMSO) [23, 24]. The relatively low viscosity of pullulan solution, salt and pH tolerance make this polymer well situated as a coating, film, and encapsulation matrix in various industrial applications [27]. It has an intermediate chemical structure between dextran and amylose. Due to the unique chemical structure, it has better structural flexibility and solubility [28].

8.2.1.1 Film formation

The powder form of pullulan is blended with appropriate quantity of water to prepare membrane, films, fibers, and compressive molding [19]. The water-insoluble pullulan films are generally prepared by either partial or complete etherification/esterification of pullulan. Pullulan solution (5–10% wt.) could be uniformly poured on the smooth surface of the plate and dried in a hot air oven at suitable temperature to make an edible thin film of 0.01 mm thickness. The formed thin film is nontoxic, transparent, thermostable, colorless, odorless, tasteless, tenacious, and resistance to oil, oxygen, and grease [19]. Moreover, the flexibility of film is significantly improved by the addition of plasticizer (up to 20%), namely, manitol, glycerol, and sorbitol [29]. Owing to better heat stability and strong resistance to both oxygen and oil, the edible thin film prepared from pullulan solution has several advantages over other polysaccharides such as starch, chitosan [30–33]. The antimicrobial and antioxidant properties of pullulan film were enhanced by the addition of other substances such as cinnamon essential oil (or) Tween 80 (or) other polysaccharides with pullulan [34].

8.2.1.2 Structural properties

Pullulan—a linear, unbranched, water-soluble α-D-glucan [35]—is produced by *A. pullulans*. It consists of repeating maltotriose units attached by α-(1→4) glycosidic bonds. The repeated maltotriose units are internally linked to each other by α-(1→6) glycosidic linkages [23, 36], as shown in Fig. 8.1.

The α-dextrorotatory index ($[\alpha]_D$) of pullulan generally ranges from +180 to +190° in an aqueous solution [19]. Researchers used FTIR and NMR spectra to study the structure of pullulan. Barker et al. reported the presence of α-glycosidic linkages between individual glycosyl residues in each maltotriose using a strong absorption peak ranging from 845 to 860 cm^{-1} [37]. The presence of (1→4)-α-D and (1→6)-α-D glycosidic linkages in pullulan was confirmed by the following transmission peak ranges; 755–795 cm^{-1} and 930–1018 cm^{-1} [19, 38]. In ^{1}H-NMR spectra, the hydroxyl (–OH) proton in the pullulan is identified by the chemical shift ranging from 4.5 to 5.6 ppm [39]. In ^{13}C NMR, 6-deoxy sugar in the sample is identified by a characteristic peak at 16.8 ppm. Moreover, the α-D-glucose in the form of pyranose in the pullulan is confirmed by the presence of signal at 92.62 ppm and absence of signals between 82 and 88 ppm [18, 40, 41]. An anomeric carbon α-(1→6) in the pullulan is characterized by a peak resonance at 100.1 ppm [40].

8.2.1.3 Molecular mass and rheological properties

Molecular mass is a crucial parameter in industrial applications. It strongly influences the gas barrier and mechanical properties (as food coating or packaging or encapsulation matrix), adhesiveness, and moisture retention (as an adhesive agent) and viscosity and

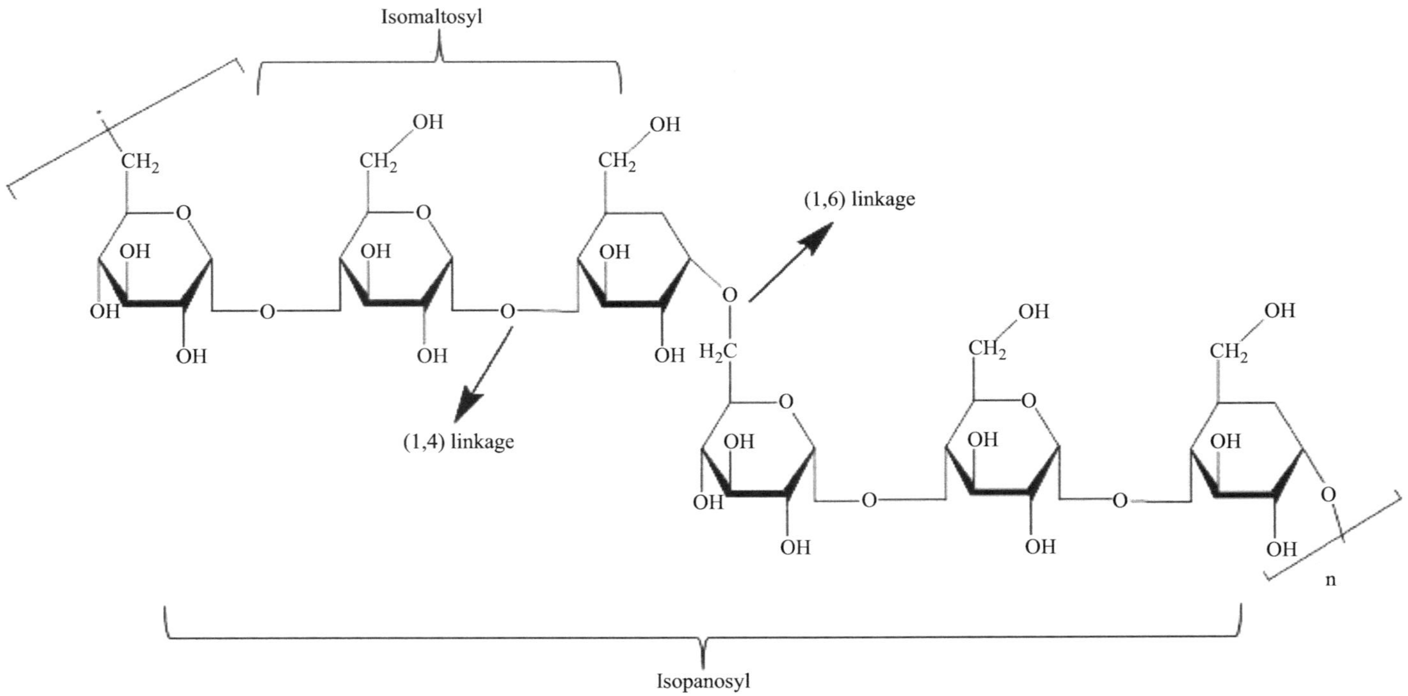

Figure 8.1 Structure of pullulan.

dispersibility (used in the preparation of low calorific food stuff). Because of biodegradability and biocompatibility, the appropriate molecular weight of pullulan can act as an immune modulator or drug carrier used in tissue engineering and pharmaceutical applications [42]. The polydispersity index (ratio of weight average molecular mass and number average molecular mass) of pullulan ranges from 2.1 to 4.1 [43]. The polydispersity index of 1.2 is preferred to be used as a blood plasma substitute in biomedical applications [44]. The average molecular mass and polydispersity index were characterized by gel permeation chromatography using standard molecular weight marker. The weight average molecular mass, number average, and degree of polymerization molecular mass and degree of polymerization of microbial pullulan are 362–480 kDa, 100–200 kDa, and 100–5000 glucopyranosyl residues, respectively [31, 45].

Carbon source [46], mode of fermentation, type of fungal strain [47], inoculum size, young active cells, composition of the production medium, initial pH [26], phosphate composition, nitrogen source, impeller speed, and age of cultivation [48] usually govern the molecular mass and size of the microbial pullulan during fermentation [49, 50]. These variables also affect the degree of polydispersity by variation in cell morphology, which is correlated with biomechanism and biosynthesis pathway [40]. During fermentation, the molecular mass of microbial pullulan gradually increases, reaches the maximum, and then gets eventually decreased due to the hydrolysis of synthesized pullulan either on the cell surface or in the bulk medium by the action of endoamylase *A. pullulans* in the later age of fermentation [51]. Usually, the high molecular mass of pullulan production was exhibited while using agriculture residue compared to synthetic carbon sources [46]. The rheological properties of the fermentation medium are greatly influenced by the molecular mass of microbial pullulan [52]. A few scientists have proposed empirical correlation relating molecular mass (M_m) with rheological property in terms of intrinsic viscosity (α), listed in Table 8.2 [53, 54].

Table 8.2 Expressions for intrinsic viscosity of pullulan

S. No.	Intrinsic viscosity (α) in cm^3/g	Molecular mass (M_m), Da	Reference
1	$\alpha = 0.0221\, M_m{}^{0.06}$	More than 1×10^5	[53]
2	$\alpha = 0.0196\, M_m{}^{0.66 \pm 0.012}$	More than 3×10^4	[54]
3	$\alpha = 0.0616\, M_m{}^{0.56 \pm 0.042}$	Less than 3×10^4	[54]

The rheological properties of the fermentation medium are greatly affected by any one of the following: (i) cell concentration; (ii) cell morphology, including shape, size, and vacuolation; (iii) deformability and flexibility of microbes; (iv) osmotic pressure of the suspending fermentation medium; (v) composition of polymeric substrate; (vi) composition of polymeric product; and (vii) shear rate. If the product of fermentation is a polymeric substances like pullulan or dextran or xanthan gum, the broth viscosity is significantly increased by the continuous excretion of polymer from cell to the fermentation medium. In this, cell concentration has negligible effect on viscosity; however, the rheological property of the broth is dominated by the production of dissolved polysaccharides during the fermentation. Because of the biosynthesis of the crosslinked network of polysaccharide (gel formation) or flow behavior index or consistency index, the polymer has the capability to change the rheological property of aqueous solution, which makes them suitable in biomedical and food industrial applications. In contrast, if the fermentation is started with polymeric substrate such as starch, there could be a progressive transformation of broth behavior from non-Newtonian to Newtonian, since apparent viscosity of broth decreases due to the consumption of polymeric substrate by the microbial system. In mycelia fermentation, even though the fermentation is started with a polymeric substance, broth behavior becomes pseudoplastic due to the development of filamentous network [55]. As compared to plant polysaccharides and other microbial polysaccharides, pullulan solutions are well known to describe Newtonian flow behavior in the wide range of polymer composition [27, 56]. This might be due to the presence of greater rotational nature of α-(1→6) glycosidic linkage yielding a random coil molecular conformation in solution [27]. The apparent viscosity and non-Newtonian index are well correlated with polymer composition and fermentation age [27, 56].

The power law equation describing the non-Newtonian behavior is given by

$$\tau = k\gamma^n$$

where τ is shear stress (N/m^2), γ is the velocity gradient (or) shear rate (s^{-1}), and k and n are consistency index and flow behavior index, respectively. If $n = 1$, then the fluid is Newtonian, otherwise non-Newtonian (pseudoplastic if $n < 1$ and dilatant if $n > 1$) [55].

Hilares et al. reported the comparison of rheological behavior between sucrose and sugarcane bagasse hydrolysate medium [57]. During the fermentation, the flow behavior index decreases, whereas the consistency index increases. This trend may be due to the pseudoplastic nature of the pullulan produced [57]. Generally, in bioprocess engineering, the fermentation medium does not exhibit dilatant behavior [55].

8.2.1.4 Thermal and mechanical properties

Thermal and mechanical properties are important factors of polysaccharides in industrial applications. The thermogravimetric property of pullulan is characterized by TGA-DSC analyzer. The decomposition point of pullulan ranges from 250 to 280°C [1]. The degree of decomposition point is affected by various factors such as molecular weight, chemical structure, nature of the sample (nonionic pullulan or chemical modification of pullulan), type of strain used, and type of fermentation [58, 59]. The molecular weight of the polymer also affects the mechanical properties such as tensile strength, compressive strength, elongation break, and tensile modules [60, 61]. These properties are significantly affected by either chemical modification of pullulan or blending of pullulan with other polysaccharides [59, 62].

8.2.2 Chemical Properties

The chemical (periodate) oxidation of pullulan yields formic acid as a by-product. The ratio of α-D-(1→6) to α-D-(1→4) glycosidic linkages in the pullulan is determined theoretically (1:2) using the quantity of production of formic acid and the consumption of periodate during the chemical oxidation [63]. Permethylation is another method to identify the glycosidic linkage ratio [64]. The operating condition and the type of hydrolysis (complete or partial) may also affect the formation of end product from pullulan. The hydrolysis of pullulan using 1 N HCl at 100°C produces monosaccharide (glucose) and oligosaccharides. Pullulan does not react with iodine solution and can form complexes with Cu^{2+} ions [65]. The heat of combustion of pullulan is approximately 16.7 kJ/g (4 kcal/g) [19]. Pullulan is a nonionic polysaccharide. The nonionic nature of pullulan has about nine –OH groups on its ring per unit of maltotriose. The chemical modifications of pullulan can be performed by various chemical reactions such as etherification, esterification, oxidation, sulfation, amidification, and copolymerization by the replacement of appropriate hydroxyl groups in the chemical structure [5, 66, 67]. In spite of the nonionic nature of polysaccharide and its applications in food and tissue engineering, the chemical modification of pullulan added its efficacy in biomedical and tissue-engineering applications due to improvement in thermal stability and mechanical property and reduction in enzymatic degradation of pullulan. The chemically modified pullulan can be used as a targeted drug-delivery material or carrier to different diseases in various human organs such as brain, lungs, spleen, and liver [68, 69]. The chemically modified pullulan, namely, pullulan acetate, pullulan carboxylate, and pullulan sulfate, is widely used as drug- or gene-delivery material in pharmaceutical industries. The industrial applications of chemically modified pullulan are discussed in Section 8.4.

8.2.3 Biochemical Properties

Pullulanase (EC.3.2.1.41, pullulan 6-glucanohydrolase) is a hydrolytic, debranching, and thermostable enzyme, which belongs to the family of glycosyl hydrolases thirteen (GH13). It is widely present in various plants, animals, and microorganisms such as bacteria and fungi. Pullulan and other amylaceous polysaccharides containing α-(1$\rightarrow$6) glycosidic linkages were hydrolyzed by this enzyme to yield maltotriose, a major product and maltotetraose, a by-product under suitable conditions [70, 71]. The mechanism of the enzymatic reaction for amylaceous type of polysaccharides or pullulan degradation and their end products are different in each case [72].

Pullulan can undergo the following enzymatic reactions by (1→6) D-pullulanase and α-(1→4) D-pullulanase: (i) hydrolysis of α-(1→6) glycosidic bond by α-(1→6) D-pullulanase, yielding a high quantity of maltotriose with trace amounts of maltotetraose; (ii) cleavage of α-(1→4) glycosidic bond by α-(1→4) D-pullulanase, which is responsible for the formation of isopanose. Due to the reaction specificity of pullulanase, it breaks specific glycosidic bonds [73].

8.3 Biosynthesis of Pullulan

In order to conserve energy, any living microorganism synthesizes substances only if they are required. Generally, polysaccharides are synthesized by microorganisms for their self-protection from chemical and environmental stress factors [74]. Biosynthesis of exopolysaccharides occurs under stressed conditions unfavorable for growth, such as osmotic, thermal stress, or for communication to other cells via biofilm formation. *A. pullulans* synthesizes pullulan to protect itself from desiccation and attacks from protozoan [75]. The mechanism for pullulan biosynthesis has not been elaborately described owing to complex cytological and physiological characteristics of *A. pullulans* [75]. Pullulan biosynthesis occurs in the cytoplasm due to the uptake of any one of the carbon sources, namely, glucose, maltose, galactose, and sucrose. At the end of the mechanism, the pullulan transport occurs through the cellular membrane, β-glucan layer to cell surface. The secreted pullulan is then freely attached with the outer cell surface to form a slime layer [76].

The biosynthesis of pullulan occurs via two major stages: (1) synthesis of precursor, uridine diphosphoglucose (UDPG) from glucose 1 phosphate using UDPG pyrophosphorylase; and (b) synthesis of pullulan via polymerization of iso-panosyl residues by the action of glucosyl transferase [75].

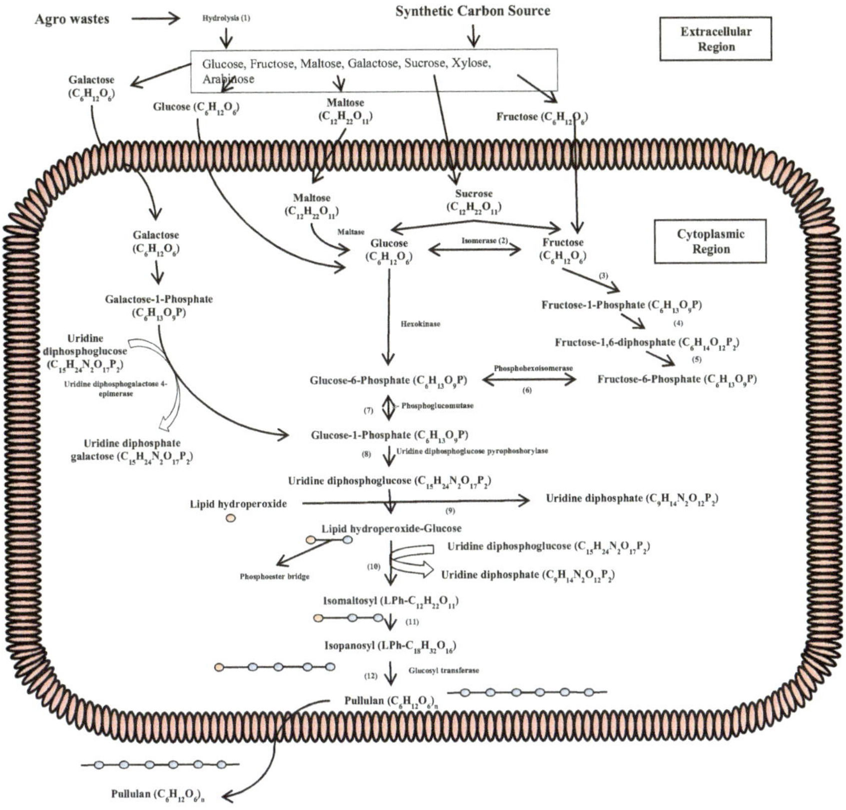

Figure 8.2 Metabolic pathway of pullulan.

The different carbon sources such as sucrose, maltose, galactose, glucose, fructose, xylose, and agrowaste are the starting raw material for the preparation of the precursor. (1) In the presence of agrowaste in the production medium, the multiple enzyme system (combination of enzymes such as amylase, protease, cellulose, xylanase, and laccase) is responsible for converting complex polymeric raw material to simple sugars. (2) Isomerization reaction takes place from glucose to fructose by the action of isomerase. (3) Then, phosphorylation reaction occurs between the synthesized fructose and ATP molecule to obtain fructose-1-phosphate. (4) With the addition of ATP molecule in the reaction mixture, fructose-1-phosphate is converted to fructose-1,6-diphosphate using fructose biphosphatase. (5) Further addition of ATP molecule with fructose-

1,6-diphosphate synthesizes fructose-6-phosphate by the action of fructose biphosphatase. (6) Then, the synthesized fructose-6-phosphate is involved in the isomerization reaction to yield glucose-6-phosphate by phosphoglucoisomerase. (7) The enzymatic conversion of glucose-6-phosphate to glucose-1-phosphate by α-phosphoglucose mutase is effectively involved in the synthesis of primary precursor, UDPG using (8) UDPG pyrophosphorylase. (9) The glucose residue from UDPG connects with lipid molecule, i.e., lipid hydroperoxide via phosphoester bridge to make lipid-linked glucosyl residue. (10) Another glucose from UDPG to the lipid-linked glucosyl residue results in the formation of isomaltosyl residue. (11) It reacts with another lipid-linked glucose to produce isopanosyl, which is the precursor for pullulan production. (12) The isopanosyl then undergoes polymerization to form pullulan. The detailed metabolic pathway is shown in Fig. 8.2.

8.3.1 Factors Influencing the Production of Pullulan in Biosynthesis

The production and yield of pullulan are greatly influenced by the nature of carbon source, type of strain, size of the inoculum, medium composition, medium pH, agitation speed, mode of fermentation (batch or fed batch or continuous), rate of aeration, type of fermentation (submerged or solid state), and nitrogen source [24, 27, 77, 78].

8.3.1.1 C-source

8.3.1.1.1 *Synthetic carbon source*

Generally, *A. pullulans* synthesizes pullulan using synthetic mono- or disaccharides, namely, xylose, glucose, fructose, mannose, galactose, sucrose, and maltose [79]. Excess carbon source in the production medium could inhibit the synthesis of pullulan during the batch cultivation [22, 80]. This might be due to the reduction in the activity of UDPG pyrophosphorylase, α-phosphoglucose mutase, and glycosyl transferase, which are responsible for the production of pullulan. Therefore, fed-batch cultivation for pullulan production is more suitable to maintain the concentration of substrate in the media within optimum range. High substrate concentration leads to the reduction in polysaccharide concentration, which is mainly due to the osmotic effects, plasmolysis, and lower level of water activity [81]. Several researchers reported the production of pullulan using synthetic carbon sources such as glucose, fructose, dextrin, maltose, and xylose. The activity of key enzymes, UDP-glucose pyrophosphorylase, α-phosphoglucose mutase, glucosyl transferase responsible for pullulan production are directly correlated with the consumption of nature of carbon source (synthetic or agrowaste) [82, 83]. The key enzyme activity was comparatively higher if the production medium contains glucose.

8.3.1.1.2 *Agrowaste residues*

Most of the researchers have focused on the production of pullulan by submerged fermentation using various hydrolysate from low-cost substrates by acid/alkali/enzymatic treatment [22, 84–87]. Agrowastes such as potato waste, sugarcane bagasse hydrolysate, molasses, sweet potato hydrolysate, brewery waste, grape pomace extract, coconut wastewater, whey, olive oil waste, rice hull hydrolysate, peat hydrolysate, carob pod, cassava bagasse, palm kernel, and jackfruit seeds are used as carbon source for pullulan production [17, 18, 21, 22, 27, 46, 57, 79, 84, 85, 88–92]. The productivity of pullulan is significantly improved using hydrolysate as raw material. However, the process cost is considerably increased owing to the hydrolysis of low-cost substrate and energy requirement in submerged fermentation [93].

8.3.1.2 N-source

Synthetic nitrogen sources such as ammonium nitrate, ammonium sulfate, urea, sodium nitrite [89, 90], and complex nitrogen sources such as soybean pomace, soybean hydrolysate, corn steep liquor, and de-oiled Jatropa seeds are widely used for the production of pullulan [86, 93, 94, 96–98]. However, several authors have used the combination of ammonium sulfate and yeast extract as nitrogen source in the production medium [40, 88, 99, 100]. Jiang et al. reported the relationship between nitrogen source and UDPG pyrophosphorylase activity [95]. The production medium containing sodium nitrate resulted higher yield of pullulan (22.46 g/L) as sodium nitrate increases the activity of UDPG pyrophosphorylase enzyme (45.07 mU/mg of protein), which is responsible for pullulan production [95]. The key enzyme activities influencing the pullulan synthesis vary with varying composition of ammonium sulfate and yeast extract. Under nitrogen-limiting condition, the activities of key enzymes were significantly changed, i.e., increase in α-phosphoglucose mutase activity; decrease in glucosyl transferase activity; and negligible change in UDPG pyrophosphorylase were observed [101].

The composition of agriculture and industrial residue may vary broadly, and very often the optimized C and N concentrations are only suitable for single substrate. To generalize the results obtained and extend the findings to a wide range of agricultural and industrial residues, C and N sources were considered a single medium variable, C/N ratio. The optimum C/N ratio can be used to design optimum medium composition irrespective of the individual carbon and nitrogen content of agriculture and industrial waste [102].

8.3.1.3 Medium pH

The yield of pullulan, activity of UDPG pyrophosphorylase, cell morphology, and molecular weight are greatly affected by medium pH [21, 103]. The optimum pH for the biosynthesis of pullulan ranges from 5.5 and 7.0 [21, 103]. The difference in optimum pH for the production of pullulan might be due to the type of fermentation, medium composition, and type of strain used. The difference in optimum pH for pullulan synthesis is reported, even though using the same microorganism, *A. pullulans*, with different subspecies. This might be due to the structure of cell membrane and wall or mass transport system of microorganism with change in cytosol pH/ osmatic stress [81]. The activity of UDPG pyrophosphorylase as well as the yield of pullulan was significantly low, if the culture medium pH is too low (<3) or too high (>7) [103]. Because of the presence of dual morphology of cell (fungal and yeast), which is responsible for product formation, several scientists reported the relationship between the yield of pullulan and cell morphology during the fermentation [16, 17, 104, 105]. The reason is that fungus-like cells and yeast-like cells were stimulated by low pH and high pH level, respectively, during the fermentation. However, the appropriate cell morphology essential for the elaboration of pullulan is still debatable. The optimum pH for cell concentration is entirely different from that for pullulan production. Thus, a two-stage pH-control strategy was effectively used for the enhancement of yield of pullulan as well as cell growth [106].

8.3.1.4 Temperature

Temperature is another crucial factor responsible for pullulan production. The optimum temperature may depend on the type of strain, fermentation medium, and culture conditions for pullulan synthesis. The difference in the optimum temperature for the production of pullulan is in the range of 24°C to 30°C and for cell growth, it is about 32°C [75, 107–111], i.e., the optimum temperature for pullulan synthesis is entirely different from that of cell growth. To improve both cell growth and polysaccharide production, a two-stage temperature-control scheme is preferred [111].

8.3.1.5 Aeration and agitation

Agitation speed describes the degree of mixing or agitation to make homogeneity or to minimize nonuniformities in the fermentation medium by eliminating the gradience of temperature, composition, and other properties [55]. A uniform or steady concentration gradient between cell and bulk medium implies the continuous transfer of medium component into the cell or secretion of product from and out of the cell through cellular membrane in a smooth way. Low shear rate causes a low rate of mass transfer, whereas high shear rate results in cell damage, which may affect the biosynthesis of pullulan during the fermentation. Therefore, agitation and aeration are essential factors affecting the production as well as molecular weight of pullulan in submerged aerobic fermentation [112, 113]. Lazaridou et al. reported the influence of agitation, aeration rate, impeller speed, and shear condition on the morphology of cell as well as biosynthesis of pullulan [114]. Strong agitation and aeration rates improve the yeast-like morphology than fungal morphology, which is essential for maximum pullulan production. But very high agitation and aeration rates cause cell damage. Similarly, Wecker and Onken reported the predominant growth of yeast-like cells, resulting in enhanced pullulan production under low shear rate conditions [115]. During the fermentation, the broth behavior becomes pseudoplastic (non-Newtonian) and more viscous due to the accumulation of pullulan in the fermentation medium. This viscous fermentation medium offers more resistance to oxygen transfer. Therefore, an appropriate combination of low shear rate and low dissolved oxygen is required to improve pullulan production as well as oxygen demand [115, 116]. Previous studies illustrated that high molecular weight as well as high concentration of pullulan could not be obtained with constant agitation speed and dissolved oxygen concentration throughout batch fermentation. However, in the two-stage agitation speed scheme, the variation in agitation speed from 500 rpm to 400 rpm significantly enhances both molecular weight and composition of pullulan simultaneously [113].

8.3.2 Purification of Pullulan

After fermentation, solid–liquid extraction (leaching) is first employed to obtain pullulan from the solid matrix using water as a solvent in solid-state fermentation [24]. The modified scheme for the recovery and purification of water-soluble polysaccharides was proposed [117]. First, the fermentation medium was subjected to centrifugation at 8000 rpm at 4°C for 40 min to remove cell debris as well as other solid residues. The supernatant obtained is allowed to precipitate by addition of three volumes of 95% (v/v) cold ethanol to the cell-free supernatant. The overnight precipitated mixture incubated at 4°C is redissolved by adding deionized water for further purification. In order to denature the cellular protein, tri-chloroacetic acid (TCA, 10%) is added and the mixture is kept at 4°C for 12 h. Then, the mixture was again centrifuged at 8000 rpm and 4°C for 20 min to remove further impurities. The supernatant collected was again blended with three volumes of 95% (v/v) cold ethanol and incubated at 4°C for 12 h. Then the sample was again dissolved in deionized water. The obtained polymer solution was dialyzed in a dialysis bag with as molecular weight cutoff of 14,000 for 48 h to remove small protein molecules and ionic matters and freeze dried for further usage, as shown in Fig. 8.3.

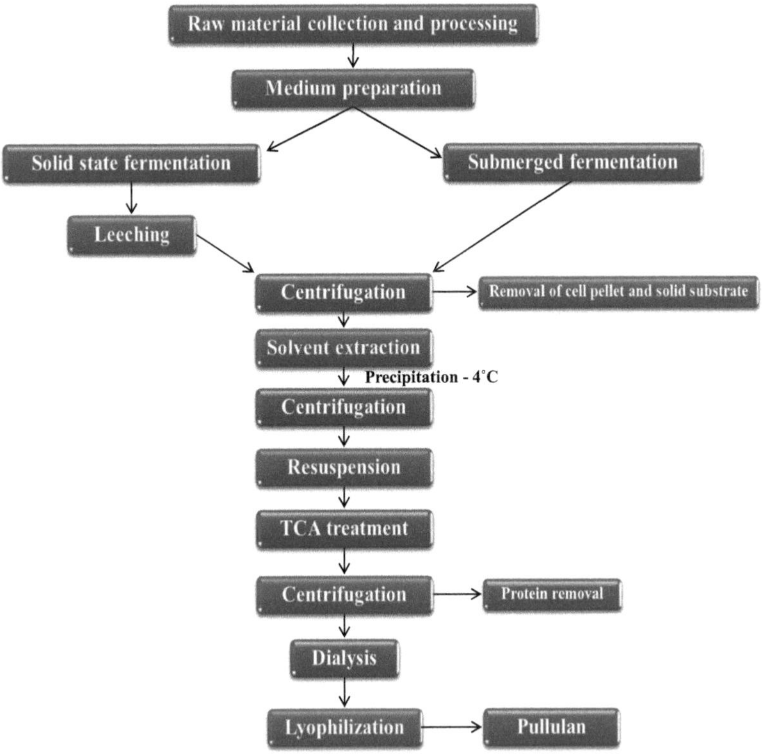

Figure 8.3 Different stages in pullulan production.

8.4 Applications of Pullulan

The chemical modification of pullulan and applications are shown in Figs. 8.4 and 8.5.

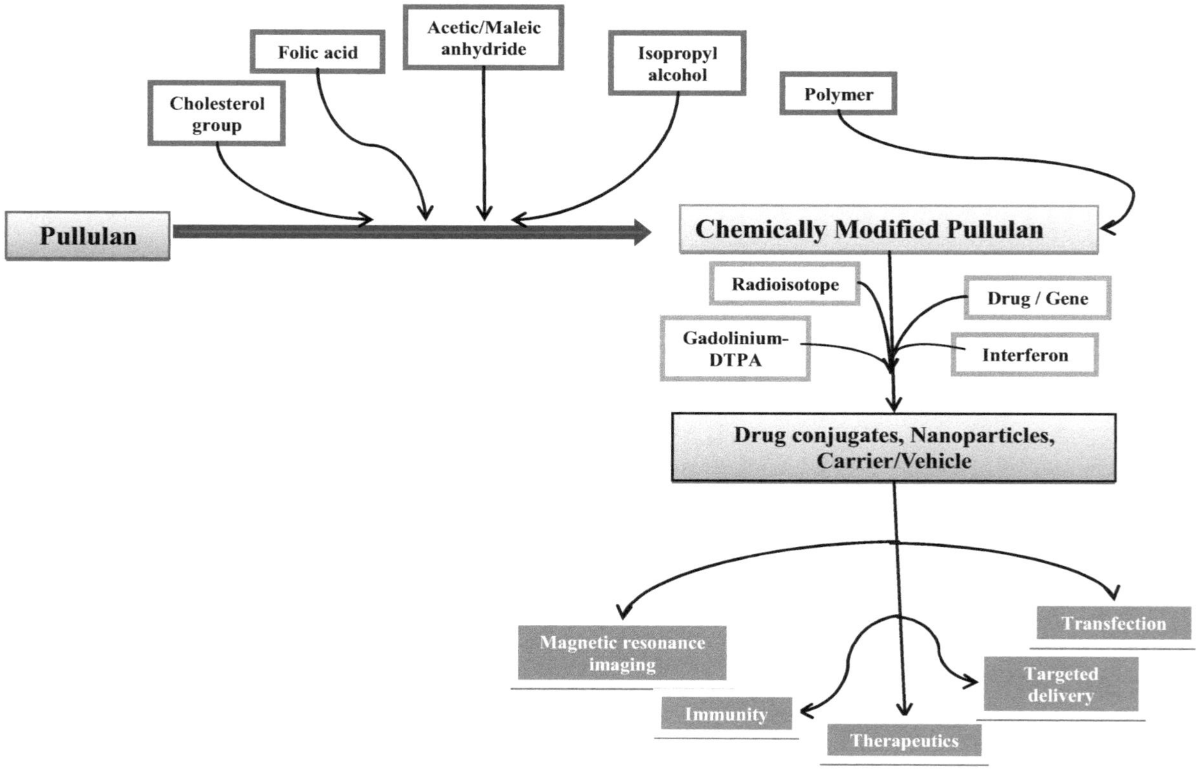

Figure 8.4 Chemical modifications of pullulan.

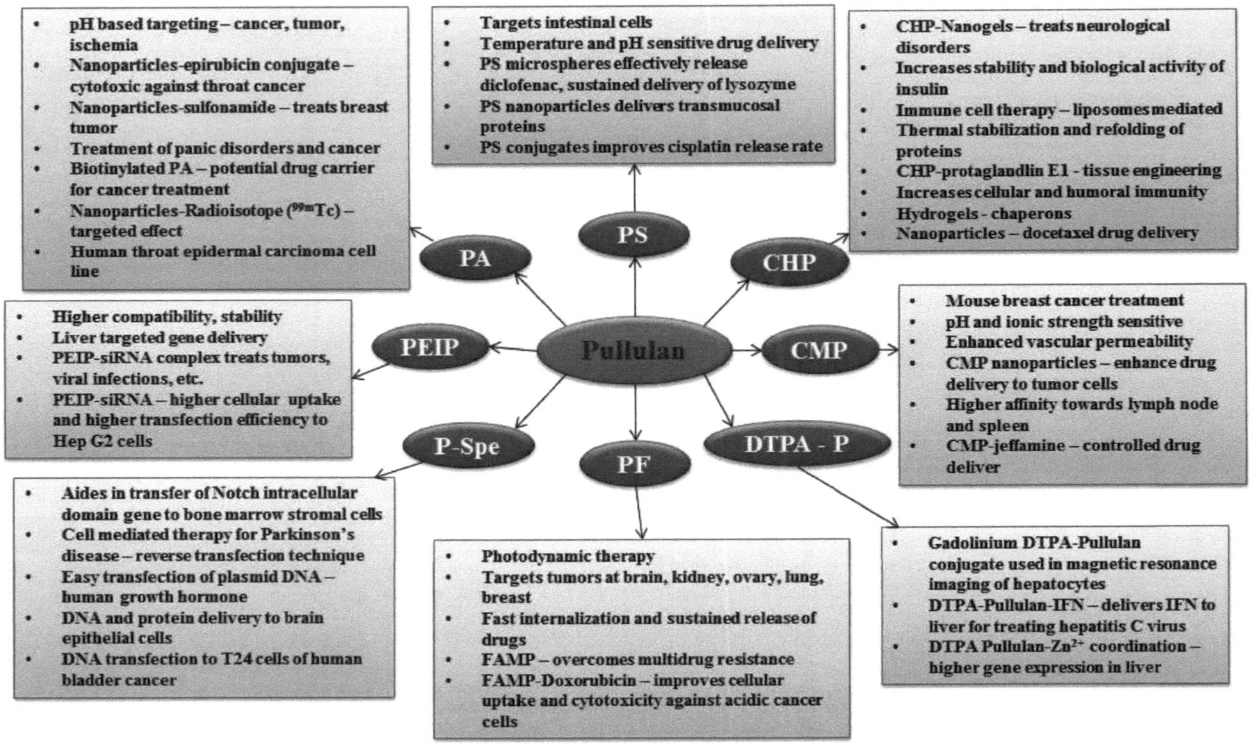

Figure 8.5 Applications of pullulan derivatives.

Abbreviations: PS: pullulan succinylate; PA: pullulan acetate; PEIP: polyethyleneimine pullulan; P-Spe: pullulan sphermine; PF: pullulan folate; DTPA-P: diethylene triamine pentaacetate dianhydride pullulan; CMP: carboxymethyl pullulan; CHP: cholesterol-bearing pullulan; siRNA: small interfering ribonucleic acid; FAMP: folate-decorated maleilated pullulan; IFN: interferon; Zn: zinc; HepG2: hepatatic G2 cells [120, 121, 128, 136, 139, 140–143, 146, 152–177].

8.4.1 Pullulan Acetate

Pullulan acetate is obtained by the acetylation of pullulan with acetic anhydride in the presence of pyridine and formamide. With the help of self-aggregation technique, the inner and outer cores of PA (less than 100 nm) are both hydrophobic and hydrophilic in nature. This specific structure of PA can act as a carrier for targeted drug delivery to treat tumor tissues or cells [118]. Jung et al. reported that an increase in concentration of drug with improved degree of acetylation causes reduction in the diffusion of drug delivery [119]. Zhang et al. reported the synthesis of PANs conjugated with Epirubicin drug containing higher cytotoxicity for the treatment of throat cancer [120]. The influence of pH on the drug-targeting PA was effectively used to treat cancer, tumor, ischemia with lower extracellular pH than surrounding cell/blood. Na et al. reported the combination of sulfonamide and PA nanoparticles for the treatment of breast tumor cell line MCF-7 [121, 122]. Biotin conjugated with PA (BPA) and vitamin H act as potential drug carrier for the treatment of cancer. Carboxymethylated polyethylene glycol (CMPEG) is added with PA to make PEP graft polymer [123]. It has been identified that PEP graft polymer has better compatibility along with the capability to influence the characteristics of hydrophobic drugs released [119]. Hydrophobic drugs such as clonazepan, silymarin encapsulated in PANPs are used in the treatment of tumor and panic disorders [119, 124]. Bishwambhar and Suneetha reported the influence of pH on naproxen-loaded PA microspheres [125]. At pH 7.2, drug release was about 75 times higher than at pH 1.2 [125]. Recently, pullulan acetate nanoparticles (PANps) were synthesized by the chemical modification of pullulan, produced by solid-state fermentation using cassava bagasse residue [102].

8.4.2 Carboxymethyl Pullulan

The addition of sodium chloroacetate and isopropyl alcohol with the main chain of pullulan forms negatively charged, high-molecular-weight, vascular permeable carboxymethyl pullulan (CMP) [126–128]. The molecular weight of negatively charged CPM is higher than nonionic native pullulan [127], and also CPM has greater vascular permeability at different pH and ionic strength [128]. As compared to free hydrophobic drug, the hydrophobic drug attached with CMP nanoparticles by hydrazone linkage improves the drug delivery to tumor cells. Lu et al. reported the high *in vitro* drug delivery at pH 5 than pH 6.4 owing to the pH sensitivity of CMP [129]. The combination of CMP with jeffamine units formed thermosensitive, anionic, and amphiphilic pullulan derivatives, Jeff M600 and Jeff M2005. Owing to this special characteristics, Jeff M600 and Jeff M2005 pullulan derivatives are extensively used in controlled drug delivery [130].

8.4.3 Cholesterol-Bearing Pullulan

Cholesterol-bearing pullulan was synthesized using the addition of cholesterol with the main chain of native pullulan by hydrophobic chemical modification. It self-aggregates in an aqueous solution to prepare spherical hydrogel nanoparticles [131–133]. CHP complexed with methacryloyl group is used to prepare nanogel. It has the specific property of chaperone, which can prevent the denaturation of enzyme, aggregation, and easily load proteins or enzymes [134, 135]. The activity of enzyme is retained (against thermal deactivation and storage) when it is complexed with CHP-CD (cyclodextrin). Similarly, several researchers have prepared CHP-CD (cyclodextrin) complex with enzyme in order to retain the activity even if the temperature is higher. The percentage activities of chymotrypsin and horse radish peroxidase were retained by 74% and 80%, respectively, even if high temperature is maintained in the enzymatic reaction [136, 137]. Nomura et al. reported the thermal stability of carbonic anhydrase B complex with CHP-CD (cyclodextrin) [138]. In the treatment of liver cancer cells and carcinoma cells, CHP is used as carrier to deliver the doxorubicin to targeted cells [139]. The stability and biological activity of insulin are improved by the addition of CHP nanogels [140]. CHP nanogels are widely used in the treatment of neurological disorders [141]. Amino-modified CHP nanoparticles are targeted drug carriers for the delivery of docetaxel [142]. CHP conjugated with physiologically active lipid component prostaglandin E1 is widely employed for neovascularization and neoepithelialization in biomedical engineering [143].

8.4.4 Pullulan–Polyetheramine

By terminating one end of the native pullulan chain using chloroformate activation, pullulan–polyetheramine can be prepared [144], and it has the capability of forming pH-sensitive and thermosensitive conjugates with drugs using reductive amination mechanism [145].

8.4.5 Polyethyleneimine Pullulan

Polyethyleneimine pullulan (PEIP) is a cationic pullulan derivative and has biocompatibility and stability in the presence of blood plasma [120, 146, 147]. The cationic nature of the synthesized pullulan derivative was investigated by adding di-ethylaminoethylamine (DEAE) or thiol group (-SH) to the native pullulan backbone chain [148, 149]. PEIP could be used as plasmid and gene carrier. PEIP–siRNA conjugate is used in the treatment of various diseases such as viral infections and tumors [120, 146]. The PEIP–siRNA complex with carboxyl groups of folate and amino groups has the capability of improving transfection with silence effect [150]. PEIP could be used as a plasmid vector carrier. Rekha and Sharma reported the transfection and binding of pGFP with hepatocytes using PEIP-plasmid vector, pGFP conjugation [151].

8.4.6 Folate-Decorated Maleilated Pullulan

The addition of nonionic pullulan with maleic anhydride leads to the synthesis of folate-decorated maleilated pullulan (FAMP) under nitrogen atmospheric condition [152]. The obtained FAMP is further treated with folic acid to make a pH-sensitive conjugate. The FAMP conjugate is used as a controlled drug-delivery material and rapid internalization of drug in various biomedical applications [153]. It is used in the treatment of tumors in various organs such as brain, ovaries, kidneys, lungs, and breast cancer cells [154]. Due to pH sensitivity and controlled drug release, doxorubicin combined with the FAMP conjugate enhances the cytotoxicity and cellular uptake by targeted acidic cancer cells [152]. This kind of conjugated nanoparticles could be employed to minimize the effect of multidrug resistance [155].

8.5 Conclusion and Future Prospects

1. Microbial pullulan, an industrially important biodegradable exopolysaccharide, is synthesized by polymorphic *A. pullulans*. Three key enzymes, namely, UDPG pyrophosphorylase, α-phosphoglucomutase, and glucosyl transferase, are involved in the biosynthesis of microbial pullulan. The factors C-source, N-source, medium pH, temperature, and agitation speed influence the activity of key enzymes for pullulan production.
2. In order to minimize the cost of fermentation, various industrial/agrowaste residues and their hydrolysates are effectively utilized for producing pullulan in solid state or submerged fermentation.
3. In the recent past, pullulan was used only in the food industry. With the help of recent advances in technology, including strain improvement, production skills, reactor design, and two-stage strategies of agitation speed/pH/temperature, pullulan production was significantly improved. Through appropriate chemical modification, pullulan derivatives have been developed with distinct characteristics of mechanical and thermal strength. The synthesized pullulan derivatives have excellent potential to serve as drug/gene carrier or vehicle in the pharma industry. In order to improve the stability and biocompatibility of pullulan for tissue-engineering applications, novel technologies for preparation of polymer–pullulan composites are focused.

References

1. Singh, R. S., Saini, G. K., and Kennedy, J. F. (2008). Pullulan: Microbial sources, production and applications, *Carbohydr. Polym.*, **73**, pp. 515–531.
2. Bounaix, M.-S., Gabriel, V. R., Morel, S., Robert, H., Rabier, P., On, M. R.-S., Gabriel, B., and Faucher, C. F. (2009). Biodiversity of exopolysaccharides produced from sucrose by sourdough lactic acid bacteria, *J. Agric. Food Chem.*, **57**, pp. 10889–10897.
3. Vijayendra, S. V. N. and Shamala, T. R. (2013). Film forming microbial biopolymers for commercial applications: A review, *Crit. Rev. Biotechnol.*, **34**(4), pp. 338–357.

4. Chawla, P. R., Bajaj, I. B., Survase, S. A., and Singhal, R. S. (2009). Microbial cellulose: Fermentative production and applications, *Food Technol. Biotechnol.*, **47**, pp. 107–124.
5. Singh, R. S. and Kaur, N. (2015). Microbial biopolymers for edible film and coating applications. In: *Advances in Industrial Biotechnology* (Nawani, N. N., Khetmalas, M., Razdan, P. N., and Pandey, A., Eds.), pp. 187–216. New Delhi, India: IK International Publishing House Pvt. Ltd.
6. Sarwat, F., Ul Qader, S. A., Aman, A., and Ahmed, N. (2008). Production and characterization of a unique dextran from an indigenous *Leuconostoc mesenteroides* CMG713, *Int. J. Biol. Sci.*, **4**(6), pp. 379–386.
7. Goulas, A. K., Corke, C. T., Grandison, A. S., and Rastall, R. A. (2004). Synthesis of isomaltooligosaccharides and oligodextrans in a recycle membrane bioreactor by the combined use of dextransucrase and dextranase, *Biotechnol. Bioeng.*, **88**(6), pp. 778–787.
8. Gunasekar, V., Reshma, K. R., Treesa, G., Gowdhaman, D., and Ponnusami, V. (2014). Xanthan from sulphuric acid treated tapioca pulp: Influence of acid concentration on xanthan fermentation, *Carbohydr. Polym.*, **102**(4), pp. 669–673.
9. USFDA. (2002). Agency Response Letter: GRAS Notice No. GRN 000099 [Pullulan]. US Food and Drug Administration (US FDA), Center for Food Safety and Applied Nutrition (CFSAN), Office of Food Additive Safety; College Park, Maryland.
10. Alban, S., Schauerte, A., and Franz, G. (2002). Anticoagulant sulfated polysaccharides: Part I. Synthesis and structure–activity relationships of new pullulan sulfates, *Carbohydr. Polym.*, **47**, pp. 267–276.
11. Sivakumar, P. A. and Rao, K. P. (2003). The use of cholesteryl pullulan for the preparation of stable vincristine liposomes, *Carbohydr. Polym.*, **51**, pp. 327–332.
12. Deshpande, M. S., Rale, V. B., and Lynch, J. M. (1992). *Aureobasidium pullulans* in applied microbiology: A status report, *Enzyme Microb. Technol.*, **14**, pp. 514–527.
13. Buliga, G. S. and Brant, D. A. (1987). Temperature and molecular weight dependence of the unperturbed dimensions of aqueous pullulan, *Int. J. Biol. Macromol.*, **9**, pp. 71–76.
14. Rekha, M. R. and Sharma, C. P. (2003). Pullulan as a promising biomaterial for biomedical applications: A perspective, *Trends Biomater. Artif. Org.*, **20**, pp. 116–121.
15. Leathers, T. D. (2003). Biotechnological production and applications of pullulan, *Appl. Microbiol. Biotechnol.*, **62**, pp. 468–473.

16. Ronen, M., Guterman, H., and Shabtai, Y. (2002). Monitoring and control of pullulan production using vision sensor, *J. Biochem. Biophys. Methods.*, **3**, pp. 243–249.

17. Sugumaran, K. R., Gowthami, E., Swathi, B., Elakkiya, S., Srivastava, S. N., Ravikumar, R., and Ponnusami, V. (2013). Production of pullulan by *Aureobasidium pullulans* from Asian palm kernel: A novel substrate, *Carbohydr. Polym.*, **92**, pp. 697–703.

18. Sugumaran, K. R., Jothy, P., and Ponnusami, V. (2014). Bioconversion of industrial solid waste-cassava bagasse for pullulan production in solid state fermentation, *Carbohydr. Polym.*, **99**, pp. 22–30.

19. LeDuy, A., Choplin, L., Zajic, J. E., and Luong, J. H. T. (2014). Pullulan: Properties, synthesis and applications. In: *Encyclopedia of Polymer Science and Technology.* John Wiley & Sons, Ltd.

20. Miller, T.L. and Churchill, B. W. (1986). Substrates for large-scale fermentations. In: *Manual of Industrial Microbiology and Biotechnology* (Demain, A. L. and Solomon, L. A., Eds.). Washington, DC: American Society for Microbiology.

21. Wu, S., Jin, Z., Tong, Q., and Chen, H. (2009). Sweet potato: A novel substrate for pullulan production by *Aureobasidium pullulans*, *Carbohydr. Polym.*, **76**, pp. 645–649.

22. Göksungur, Y., Uzunoğullari, P., and Dağbağli, S. (2011). Optimization of pullulan production from hydrolysed potato starch waste by response surface methodology, *Carbohydr. Polym.*, **83**, pp. 1330–1337.

23. Singh, R. S., Kaur, N., and Kennedy, J. F. (2019). Pullulan production from agro-industrial waste and its applications in food industry: A review, *Carbohydr. Polym.*, **217**, pp. 46–57.

24. Sugumaran, K. R. and Ponnusami, V. (2017). Review on production, downstream processing and characterization of microbial pullulan, *Carbohydr. Polym.*, **173**, pp. 573–591.

25. Teekamp, N., Tian, Y., Visser, J. C., Olinga, P., Frijlink, H. W., Woerdenbag, H. J., and Hinrichs, W. L. J. (2017). Addition of pullulan to trehalose glasses improves the stability of β-galactosidase at high moisture conditions, *Carb. Polym.*, **176**, pp. 374–380.

26. Lee, K. Y. and Yoo, Y. J. (1993). Optimization of pH for high molecular weight pullulan, *Biotechnol. Lett.*, **15**, pp. 1021–1024.

27. Lazaridou, A., Biliaderis, C. G., Roukas, T., and Izydirczyk, M. (2002). Production and characterization of pullulan from beet molasses using a nonpigmented strain of *Aureobasidium pullulans* in batch culture, *Appl. Biochem. Biotechnol.*, **97**, pp. 1–22.

28. Gniewosz, M. and Duszkiewicz-Reinhard, W. (2008). Comparative studies on pullulan synthesis, melanin synthesis and morphology of white mutant *Aureobasidium pullulans* B-1 and parent strain A. P-3, *Carbohydr. Polym.*, **72**, pp. 431–438.
29. Bender, H., Lehmann, J., and Wallenfels, K. (1959). Pullulan, an extracellular glucan from *Pullularia pullulans, Biochem. Biophys. Acta*, **36**, 309.
30. Diab, T., Biliaderis, C. G., Gerasopoulos, D., and Sfakiotakis, E. (2001). Physico-chemical properties and application of pullulan edible films and coatings in fruit preservation, *J. Sci. Food Agric.*, **81**(10), pp. 988–1000.
31. Yuen, S. (1974). Pullulan and its applications, *Process Biochem*., **9**, pp. 7–22.
32. Wu, J., Zhong, F., Li, Y., Shoemaker, C. F., and Xia, W. (2013). Preparation and characterization of pullulan–chitosan and pullulan–carboxymethyl chitosan blended films, *Food Hydrocoll.*, **30**(1), pp. 82–91.
33. Farris, S., Unalan, I. U., Introzzi, L., Fuentes-Alventosa, J. M., and Cozzolino, C. A. (2014). Pullulan-based films and coatings for food packaging: Present applications, emerging opportunities, and future challenges, *J. Appl. Polym. Sci.*, **131**(13).
34. Chu, Y., Xu, T., Gao, C. C., Liu, X., Zhang, N., Feng, X., Liu, X., Shen, X., and Tang, X. (2019). Evaluations of physicochemical and biological properties of pullulan-based films incorporated with cinnamon essential oil and Tween 80, *Int. J. Biol. Macromol.*, **122**, pp. 388–394.
35. Carolan, G., Catley, B. J., and McDougal, F. G. (1983). The location of tetrasaccharide units in pullulan, *Carbohydr. Res.*, **114**, pp. 237–243.
36. Sutherland, L. W. (1998). Novel and established applications of microbial polysaccharides, *Trends Biotechnol.*, **16**, pp. 41–46.
37. Barker, S. A., Bourne, E. J., Stacey, M., and Whiffen, D. H. (1954). Infrared spectra of carbohydrates. Part I. Some derivatives of D-glucopyranose, *J. Chem. Soc.*, **75**, pp. 171–176.
38. Karim, M. R. and Islam, S. (2011). Thermal behavior with mechanical property of fluorinated silane functionalized super hydrophobic pullulan / poly (vinyl alcohol) blends by electrospinning method, *J. Nanomater.*, Article ID: 979458.
39. Zhang, H.-Z., Gao, F.-P., Liu, L.-R., Li, X.-M., Zhou, Z.-M., Yang, X.-D., and Zhang, Q.-Q. (2009). Pullulan acetate nanoparticles prepared by solvent diffusion method for Epirubicin chemotherapy, *Colloids Surf. B.*, **71**, pp. 19–26.

40. Singh, R. S., Saini, G. K., and Kennedy, J. F. (2009). Downstream processing and characterization of pullulan from a novel color variant strain of *Aureobasidium pullulans* FB-1. *Carbohydr. Polym.*, **78**, pp. 89–94.

41. Ye, L., Zhang, J., Ye, X., Tang, Q., Liu, Y., Gong, C., Du, X., and Pan, Y. (2008). Structural elucidation of the polysaccharide moiety of a glycopeptide (GLPCW-II) from *Ganoderma lucidum* fruiting bodies, *Carbohydr. Res.*, **343**(4), pp. 746–752.

42. Seibutsu, H. and Kenkyujo, M. (1983). Hayashibara Biochemical Laboratories, assignee. UK patent GB 2,109,391.

43. Wiley, B. J., Ball, D. H., Arcidiacono, S. M., and Kaplan, D. L. (1993). Control of molecular weight distribution of the biopolymer pullulan produced by *Aureobasidium pullulans*, *J. Polym. Environ.*, **1**, pp. 3–9.

44. Singh, R. S., Kaura, N., and Kennedy, J. F. (2015). Pullulan and pullulan derivatives as promising biomolecules for drug and gene targeting, *Carbohydr. Polym.*, **123**, pp. 190–207.

45. Okada, K., Yoneyama, M., Mandai, T., Aga, H., Sakai, S., and Ichikawa, T. (1990). Digestion and fermentation of pullulan, *J. Japan Soc. Nutri. Food Sci.*, **43**, pp. 23–29.

46. Israilides, C., Scanlon, B., Smith, A., Harding, S. E., and Jumel, K. (1994). Characterization of pullulans produced from agro-industrial wastes, *Carbohydr. Polym.*, **25**, pp. 203–209.

47. Pollock, T. J., Thorne, L., and Armentrout, R. W. (1992). Isolation of new *Aureobasidium* strains that produce high-molecular weight pullulan with reduced pigmentation, *Appl. Environ. Microbiol.*, **58**, pp. 877–883.

48. McNeil, B. and Kristiansen, B. (1987). Influence of impeller speed upon the pullulan fermentation, *Biotechnol. Lett.*, **9**, pp. 101–104.

49. Kim, J. H., Kim, M. R., Lee, J. H., Lee, J. W., and Kim S. K. (2000). Production of high molecular weight pullulan by *Aureobasidium pullulans* using glucosamine, *Biotechnol. Lett.*, **22**, pp. 987–990.

50. Lee, J. H., Kim, J. H., Zhu, I. H., Zhan, X. B., Lee, J. W., Shin, D. H., and Kim, S. K. (2001). Optimization of conditions for the production of pullulan and high molecular weight pullulan by *Aureobasidium pullulans*, *Biotechnol. Lett.*, **23**(10), pp. 817–820.

51. West, T. P. and Strohfus, B. (1996). A pullulan-degrading enzyme activity of *Aureobasidium pullulans*, *J. Basic Microbiol.*, **36**, pp. 377–380.

52. Buliga, G. S. and Brant, D. A. (1987). Temperature and molecular weight dependence of the unperturbed dimensions of aqueous pullulan, *Int. J. Biol. Macromol.*, **9**, pp. 71–76.

53. Kato, T., Katsuki, T., and Takahashi, A. (1984). Static and dynamic solution properties of pullulan in a dilute solution, *Macromolecules*, **17**, pp. 1726–1730.

54. Tsvetkov, V. N. (1986). *Rigid-Chain Polymers.* New York: Plenum.

55. Doran, P. (2012). *Bioprocess Engineering Principles*, 2nd Edition. Academic Press.

56. LeDuy, A., Marsan, A. A., and Coupal, B. (1974). A study of the rheological properties of a non-Newtonian fermentation broth, *Biotechnol. Bioeng.*, **16**, pp. 61.

57. Hilares, R. T., Resende, J., Orsi, A. C., Ahmed, M. A., Lacerda, M. T., da Silva, S. S., and Santos, J. C. (2019). Exopolysaccharide (pullulan) production from sugarcane bagasse hydrolysate aiming to favor the development of biorefineries, *Int. J. Bio. Macro.*, **127**, pp. 169–177.

58. Srikanth, S., Swathi, M., Tejaswini, M., Sharmila, G., Muthukumaran, C., Jaganathan, M. K., and Tamilarasan, K. (2014). Statistical optimization of molasses-based exopolysaccharide and biomass production by *Aureobasidium pullulans* MTCC 2195, *Biocat. Agri. Biotechnol.*, **3**(3), pp. 7–12.

59. Teramoto, N. and Shibata, M. (2006). Synthesis and properties of pullulan acetate: Thermal properties, biodegradability, and a semi-clear gel formation in organic solvents, *Carbohydr. Polym.*, **63**, pp. 476–481.

60. Ito, M., Takahashi, A., Araki, N., and Kanamoto, T. (2001). Effects of molecular weight on the drawing and the draw efficiency of nylon-6, *Polymer*, **42**, pp. 241–248.

61. Jordens, K., Wilkes, G. L., Janzen, J., Rohlfing, D. C., and Welch, M. B. (2000). The influence of molecular weight and thermal history on the thermal, rheological, and mechanical properties of metallocene-catalyzed linear polyethylenes, *Polymer*, **41**, pp. 7175–7192.

62. Trovatti, E. and Fernandes, S. C. M. (2012). Sustainable nanocomposite films based on bacterial cellulose and pullulan, *Cellulose*, **19**, pp. 729–737.

63. Jackson, E. L. and Hudson, C. S. (1937). Application of the cleavage type of oxidation by periodic acid to starch and cellulose, *J. Am. Chem. Soc.*, **59**, pp. 994.

64. Wallenfels, K., Bechtler, G., Kuhn, R., Trischmann, H., and Egge, H. (1963). Analytisch-technische unter-suchungen, *Angew. Chem.*, **75**, pp. 1014.

65. Ueda, S., Fujita, K., Komatsu, K., and Nakashima, Z. (1963). Polysaccharide produced by the genus *Pullularia.* I. Production of polysaccharide by growing cells, *Appl. Microbiol.*, **11**, pp. 211–215.

66. Shingel, K. I. (2004). Current knowledge on biosynthesis, biological activity, and chemical modification of the exopolysaccharide, pullulan, *Carbohdr. Res.*, **339**, pp. 447–460.

67. Bataille, I., Meddahi-Pellé, A., Le Visage, C., Letourneur, D., and Chaubet, F. (2011). Pullulan for biomedical uses. In: *Polysaccharides in Medicinal and Pharmaceutical Applications* (Popa, V., Ed.), pp. 145–182. UK: iSmithers Rapra Publishing.

68. Prajapati, V. D., Jani, G. K., and Khanda, S. M. (2013). Pullulan: An exopolysaccharide and its various applications, *Carbohydr. Polym.*, **95**, pp. 540–549.

69. Singh, R. S. and Saini, G. K. (2014). Pullulan as therapeutic tool in biomedical applications. In: *Advances in Industrial Biotechnology* (Singh, R. S., Pandey, A., and Larroche, C., Eds.), pp. 263–291. India: IK International Publishing House Pvt. Ltd.

70. Janecek, S., Svensson, B., and Henrissat, B. (1997). Domain evolution in the α-amylase family, *J. Mol. Evol.*, **45**, pp. 322–331.

71. Matzke, J., Herrmann, A., Schneider, E., and Bakker, E. P. (2000). Gene cloning, nucleotide sequence and biochemical properties of a cytoplasmic cyclomaltodextrinase (neopullulanase) from *Alicyclobacillus acidocaldarius*, reclassification of a group of enzymes, *FEMS Microbiol. Let.*, **183**, pp. 55–61.

72. Domań-Pytka, M. and Bardowski, J. (2004). Pullulan degrading enzymes of bacterial origin, *Crit. Rev. Microbiol.*, **30**, pp. 107–121.

73. Singh, R. S., Saini, G. K., and Kennedey, J. F. (2011). Continuous hydrolysis of pullulan using covalently immobilized pullulanase in a packed bed reactor, *Carbohydr. Polym.*, **83**(2), pp. 672–675.

74. Ponnusami, V. and Gunasekar, V. (2015). *Production of Pullulan by Microbial Fermentation*. Switzerland: Springer International Publishing.

75. Cheng, K. C., Demirci, A., and Catchmark, J. M. (2011). Pullulan: Biosynthesis, production and applications, *Appl. Microbiol. Biotechnol.*, **92**, pp. 29–44.

76. Simon, L., Caye-Vaugien, C., and Bouchonneau, M. (1993). Relation between pullulan production, morphological state and growth conditions in *Aureobasidium pullulans*: New observations, *J. Gen. Microbiol.*, **139**, pp. 979–985.

77. Campbell, S. B., McDougall, M. B., and Seviour, J. R. (2003). Why do exopolysaccharide yields from the fungus *Aureobasidium pullulans* fall during batch culture fermentation, *Enzyme Microb. Tech.*, **33**, pp. 104–112.

78. Sugumaran, K. R. and Ponnusami, V. (2017). Conventional optimization of aqueous extraction of pullulan in solid-state fermentation of cassava bagasse and Asian palm kernel, *Biocatal. Agri. Biotechnol.*, **10**, pp. 204–208.

79. Thirumavalavan, K., Manikkandan, T. R., and Dhanasekar, R. (2009). Pullulan production from coconut by-products by *Aureobasidium pullulans, Afr. J. Biotechnol.*, **8**, pp. 254–258.

80. Cheng, K. C., Demirci, A., and Catchmark, J. M. (2010b). Effects of plastic composite support and pH profiles on pullulan production in a biofilm reactor, *Appl. Microbiol. Biotechnol.*, **86**(3), pp. 853–861.

81. Singh, R., Gaur, R., Tiwari, S., and Gaur, M. K. (2012). Production of pullulan by a thermotolerant *Aureobasidium pullulans* strain in non-stirred fed batch fermentation process, *Braz. J. Microbiol.*, **43**, pp. 1042–1050.

82. Duan, X. H., Chi, Z. M., Wang, L., and Wang, X. H. (2008). Influence of different sugars on pullulan production and activities of α-phosphoglucose mutase, UDPG-pyrophosphorylase and glucosyltransferase involved in pullulan synthesis in *Aureobasidium pullulans* Y68, *Carbohydr. Polym.*, **73**, pp. 587–593.

83. Grobben, G. J., Smith, M. R., Sikkema, J., and de Bont, A. M. (1996). Influence of fructose and glucose on the production of exopolysaccharides and the activities of enzymes involved in the sugar metabolism and the synthesis of sugar nucleotides in *Lactobacillus delbrueckii* subsp. bulgaricus NCFB 2772, *Appl. Microbiol. Biotechnol.*, **46**(3), pp. 279–284.

84. Roukas, T. (1998). Pretreatment of beet molasses to increase pullulan production, *Process Biochem.*, **33**, pp. 805–810.

85. Roukas, T. C. (1999). Pullulan production from brewery wastes by *Aureobasidium pullulans, World J. Microbiol. Biotechnol.*, **15**, pp. 447–450.

86. Seo, H. P., Son, C. W., Chung, C. H., Jung, D., Kim, S. K. and Gross, R. A. (2004). Production of high molecular weight pullulan by

Aureobasidium pullulans HP-2001 with soybean pomace as a nitrogen source, *Bioresour. Technol.*, **95**, pp. 292–299.

87. Chen, Y., Guo, J., Li, F., Liu, M., Zhang, X., Guo, X., and Xiao, D. (2014). Production of pullulan from xylose and hemicellulose hydrolysate by *Aureobasidium pullulans* AY82 with pH control and DL-dithiothreitol: Addition, *Biotechnol. Bioproc. E.*, **19**, pp. 282–288.

88. Wang, D., Ju, X., Zhou, D., and Wei, G. (2014). Efficient production of pullulan using rice hull hydrolysate by adaptive laboratory evolution of *Aureobasidium pullulans*, *Bioresour. Technol.*, **164**, pp. 12–19.

89. LeDuy, A. and Boa, J. M. (1983). Pullulan production from peat hydrolysate, *Can. J. Microbiol.*, **29**, pp. 143–146.

90. Roukas, T. (1999b). Pullulan production from deproteinized whey by *Aureobasidium pullulans*, *J. Ind. Microbiol. Biot.*, **22**, pp. 617–621.

91. Roukas, T. and Biliaderis, C. G. (1995). Evaluation of carob pod as a substrate for pullulan production by *Aureobasidium pullulans*, *Appl. Biochem. Biotechnol.*, **55**, pp. 27–44.

92. Singh, R. S. and Kaur, N. (2019). Understanding response surface optimization of medium composition for pullulan production from de-oiled rice bran by *Aureobasidium pullulans*, *Food Sci. Biotechnol.*, **28**(5), pp. 1507–1520.

93. Sharma, N., Prasad, G. S., and Choudhury, A. R. (2013). Utilization of corn steep liquor for biosynthesis of pullulan: An important exopolysaccharide, *Cabrohydr. Polym.*, **93**, pp. 95–101.

94. Auer, D. P. F. and Seviour, R. J. (1990). Influence of varying nitrogen sources on polysaccharide production by *Aureobasidium pullulans* in batch culture, *Appl. Microbiol. Biotechnol.*, **32**, pp. 637–644.

95. Jiang, L., Wu, S., and Kim, J. M. (2011). Effect of different nitrogen sources on activities of UDPG-pyrophosphorylase involved in pullulan synthesis and pullulan production by *Aureobasidium pullulans*, *Carbohydr. Polym.*, **86**, pp. 1085–1088.

96. Chi, Z. and Zhao, S. (2003). Optimization of medium and new cultivation conditions for pullulan production by a new pullulan-producing yeast strain, *Enzyme Microb. Technol.*, **33**, pp. 206–211.

97. Mehta, A., Prasad, G. S., and Choudhury, A. R. (2014). Cost effective production of pullulan from agri-industrial residues using response surface methodology, *Int. J. Biol. Macromol.*, **64**, pp. 252–256.

98. West, T. P. and Reed-Hamer, B. (1991). Ability of *Aureobasidium pullulans* to synthesize pullulan upon selected sources of carbon and nitrogen, *Microbios*, **67**, pp. 117–124.

99. Goksungur, Y., Dagbagli, S., Ucan, A., and Guvenc, U. (2005). Optimization of pullulan production from synthetic medium by *Aureobasidium pullulans* in a stirred tank reactor by response surface methodology, *J. Chem. Technol. Biotechnol.*, **80**, pp. 819–827.

100. Wang, D., Yu, X., and Gongyuan, W. (2013). Pullulan production and physiological characteristics of *Aureobasidium pullulans* under acid stress, *Appl. Microbilol. Biotechnol.*, **97**, pp. 8069–8077.

101. Wang, D., Chen, F., Wei, G., Jiang, M., and Dong, M. (2015). The mechanism of improved pullulan production by nitrogen limitation in batch culture of *Aureobasidium pullulans, Carb. Polym.*, **127**, pp. 325–331.

102. Sugumaran, K. R. and Ponnusami, V. (2015). Statistical modeling of pullulan production and its application in pullulan acetate nanoparticles synthesis, *Int. J. Biol. Macromol.*, **81**, pp. 867–876.

103. Pan, S., Yao, D., Chen, J., and Wu, S. (2013). Influence of controlled pH on the activity of UDPG-pyrophosphorylase in *Aureobasidium pullulans, Carbohydr. Polym.*, **92**, pp. 629–632.

104. Campbell, B. S., Siddique, A. B., McDougall, B. M., and Seviour, R. J. (2004). Which morphological forms of the fungus *Aureobasidium pullulans* are responsible for pullulan production, *FEMS Microbiol. Lett.*, **232**, pp. 225–228.

105. Li, B. X., Zhang, N., Peng, Q., Yin, T., Guan, F. F., Wang, G. L., and Li, Y. (2008). Production of pigment-free pullulan by swollen cell in *Aureobasidium pullulans* NG which cell differentiation was affected by pH and nutrition, *Appl. Microbiol. Biotechnol.*, **84**, pp. 293–300.

106. Lacroix, C., LeDuy, A., Noel, G., and Choplin, L. (1985). Effect of pH on the batch fermentation of pullulan from sucrose medium, *Biotechnol. Bioeng.*, **27**, pp. 202–207.

107. McNeil, B. and Kristiansen, B. (1990). Temperature effects on polysaccharide formation by *Aureobasidium pullulans* in stirred tanks, *Enzyme Microbial. Technol.*, **12**, pp. 521–526.

108. West, T. P. and Reed-Hamer, B. (1993). Effect of temperature on pullulan production in relation to carbon source, *Microbios*, **75**, pp. 261–268.

109. Gibson, L. H. and Coughlin, R. W. (2002). Optimization of high molecular weight pullulan production by *Aureobasidium pullulans* in batch fermentations, *Biotechnol. Prog.*, **18**, pp. 675–678.

110. Prasongsuk, S., Berhow, M. A., Dunlap, C. A., Weisleder, D., Leathers, T. D., Eveleigh, D. E., and Punnapayak, H. (2007). Pullulan production by tropical isolates of *Aureobasidium pullulans, J. Ind. Microbiol. Biotechnol.*, **34**, pp. 55–61.

111. Wu, S., Chen, H., Jin, Z., and Tong, Q. (2010). Effect of two-stage temperature on pullulan production by *Aureobasidium pullulans*, *World J. Microbiol. Biotechnol.*, **26**, pp. 737–741.

112. Survase, S. A., Saudagar, P. S., Bajaj, I. B., and Singhal, R. S. (2007). Schleroglucan: Fermentative production, downstream processing and applications, *Food Technol. Biotechnol.*, **45**, pp. 107–118.

113. Wang, D., Bian, J., Wei, G., Jiang, M., and Donga, M. (2014). Simultaneously enhanced production and molecular weight of pullulan using a two-stage agitation speed control strategy, *J. Chem. Technol. Biotechnol.*, DOI:10.1002/jctb.4600.

114. Lazaridou, A., Roukas, T., Biliaderis, C. G., and Vaikousi, H. (2002). Characterization of pullulan produced from beet molasses by *Aureobasidium pullulans* in a stirred tank reactor under varying agitation, *Enzyme Microb. Technol.*, **31**, pp. 122–132.

115. Wecker, A. and Onken, U. (1991). Influence of dissolved oxygen concentration and shear rate on the production of pullulan by *Aureobasidium pullulans*, *Biotechnol. Lett.*, **13**, pp. 155–160.

116. Seviour, R. J., Stasinopoulos, S. J., Auer, D. P. F., and Gibbs, P. A. (1992). Production of pullulan and other exopolysaccharides by filamentous fungi, *Crit. Rev. Biotechnol.*, **12**, pp. 279–298.

117. Guangbin, Y., Genliang, L., Changli, W., Ling, B., Ruirui, Y., and Suoyi, H. (2019). Extraction and characterization of dextran from *Leuconostoc pseudomesenteroides* YB-2 isolated from mango juice, *Carb. Polym.*, **207**, pp. 218–223.

118. Yokoyama, M., Kwon, G. S., Okano, T., Sakurai, Y., and Kataoka, K. (1994). Development of micelle-forming polymeric drug with superior anticancer activity. In: *Polymeric Drugs and Drug Administration* (Ottenbrite, R. M., Ed.), p. 126. Washington, DC: American Chemical Society.

119. Jung, S.-W., Jeong, Y. I., and Kim, S.-H. (2003). Characterization of hydrophobized pullulan with various hydrophobicities, *Int. J. Pharm.*, **254**, pp. 109–121.

120. Zhang, N., Tan, C., Cai, P., Zhang, P., Zhao, Y., and Jiang, Y. (2009). RNA interference in mammalian cells by siRNAs modified with morpholino nucleoside analogues, *Bioorg. Med. Chem.*, **17**, pp. 2441–2446.

121. Na, K., Shin, D., Yun, K., Park, K.-H., and Lee, K. C. (2003). Conjugation of heparin into carboxylated pullulan derivatives as an extracellular matrix for endothelial cell culture, *Biotechnol. Lett.*, **25**, pp. 381–385.

122. Na, K., Park, K. M., Jo, E. A., and Lee, K. S. (2006). Self-organized pullulan/deoxycholicacid nanogels: Physicochemical characterization

and anti-cancer drug-releasing behavior, *Biotechnol. Bioproc. E.*, **11**, pp. 262–267.

123. Jung, S.-W., Jeong, Y.-L., Kim, Y.-H., and Kim, S.-H. (2004). Self-assembled polymeric nanoparticles of poly(ethylene glycol) grafted pullulan acetate as a novel drug carrier, *Arch. Pharm. Res.*, **27**, pp. 562–569.

124. Kumar, B. S., Kumar, M. G., Suguna, L., Sastry, T. P., and Mandal, A. B. (2012). Pullulan acetate nanoparticles based delivery system for hydrophobic drug, *Int. J. Pharm. Bio. Sci.*, **3**, pp. 24–32.

125. Bishwambhar, M. and Suneetha, V. (2012). Release study of naproxen, a modern drug from pH sensitive pullulan acetate microsphere, *Intl. J. Drug Dev. Res.*, **4**, pp. 184–191.

126. Dulong, V., Mocanu, G., and LeCerf, D. (2007). A novel amphiphilic pH-sensitive hydrogel based on pullulan, *Colloid Polym. Sci.*, **285**, pp. 1085–1091.

127. Nogusa, H., Yano, T., Okuno, S., Hamana, H., and Inoue, K. (1995). Synthesis of carboxymethyl pullulan-peptide-doxorubicin conjugates and their properties, *Chem. Pharm. Bull.*, **43**, pp. 1931–1936.

128. Takakura, Y. and Hashida, M. (1996). Macromolecule carrier systems for targeted drug delivery: Pharmacokinetic consideration on for biodistribution, *Pharm. Res.*, **13**, pp. 820–831.

129. Lu, D. X., Wen, X.-T., Liang, J., Zhang, X.-D., Gu, Z.-W., and Fan, Y.-J. (2008). Novel pH sensitive drug delivery system based on natural polysaccharide for doxorubicin release, *Chin. J. Polym. Sci.*, **26**, pp. 369–374.

130. Mocanu, G., Mihai, D., Dulong, V., Picton, L., and LeCerf, D. (2011). New anionic amphiphilic thermosensitive pullulan derivatives, *Carbohydr. Polym.*, **84**, pp. 276–281.

131. Akiyoshi, K. and Sunamoto, J. (1996). Supramolecular assembly of hydrophobized polysaccharides, *Supramol. Sci.*, **3**, pp. 157–163.

132. Akiyoshi, K., Deguchi, S., Moriguchi, N., Yamaguchi, S., and Sunamoto, J. (1993). Self-aggregates of hydrophobized polysaccharides in water. Formation and characteristic of nanoparticles, *Macromolecules*, **26**, pp. 3062–3068.

133. Akiyoshi, K., Deguchi, S., Tajma, H., Nishikawa, T., and Sunamoto, J. (1997). Microscopic structure and thermo-responsiveness of a hydrogel nanoparticle by self-assembly of a hydrophobized polysaccharide, *Macromolecules*, **30**, pp. 857–861.

134. Morimoto, N., Endo, T., Iwasaki, Y., and Akiyoshi, K. (2005). Design of hybrid hydrogels with self-assembled nanogels as cross-

linkers: Interaction with proteins and chaperone-like activity, *Biomacromolecules,* **6**, pp. 1829–1834.

135. Morimoto, N., Endo, T., Ohtomi, M., Iwasaki, Y., and Akiyoshi, K. (2005). Hybrid nanogels with physical and chemical cross-linking structures as nanocarriers, *Macromol. Biosci.*, **5**, pp. 710.

136. Nishikawa, T., Akiyoshi, K., and Sunamoto, J. (1994). Supramolecular assembly between nanoparticles of hydrophobized polysaccharide and soluble protein complexation between the self-aggregate of cholesterol-bearing pullulan and α-chymotrypsin, *Macromolecules,* **27**, pp. 7654–7659.

137. Sawada, S.-I., Sasaki, Y., Nomura, Y., and Akiyoshi, K. (2011). Cyclodextrin-responsive nanogel as an artificial chaperone for horseradish peroxidase, *Colloid Polym. Sci.*, **289**, pp. 685–691.

138. Nomura, Y., Sasaki, Y., Takagi, M., Narita, T., Aoyama, Y., and Akiyoshi, K. (2005). Thermoresponsive controlled association of protein with a dynamic nanogel of hydrophobized polysaccharide and cyclodextrin: Heat shock protein-like activity of artificial molecular chaperone, *Biomacromolecules*, **6**, pp. 447–452.

139. Li, H., Bian, S., Huang, Y., Liang, J., Fan, Y., and Zhang, X. (2014). High drug loading pH-sensitive pullulan–DOX conjugate nanoparticles for hepatic targeting, *J. Biomed. Mater. Res. A*, **102A**, pp. 150–159.

140. Akiyoshi, K., Kobayashi, S., Shichibe, S., Mix, D., Baudys, M., Kim, S. W., and Sunamoto, J. (1998). Self-assembled hydrogel nanoparticle of cholesterol-bearing pullulan as a carrier of protein drugs: Complexation and stabilization of insulin, *J. Control. Release,* **54**, pp. 313–320.

141. Boridy, S., Takahashi, H., Akiyoshi, K., and Maysinger, D. (2009). The binding of pullulan modified cholesteryl nanogels to Aβ oligomers and their suppression of cytotoxicity, *Biomaterials*, **30**, pp. 5583–5591.

142. Satoh, K., Chen, F., Aoyama, A., Date, H., and Akiyoshi, K. (2008). Nanoparticle of cholesterol-bearing pullulan as a carrier of anticancer drugs, *Eur. J. Cancer Supplements*, **6**, pp. 139.

143. Kobayashi, H., Katakura, O., Morimoto, N., Akiyoshi, K., and Kasugai, S. (2009). Effects of cholesterol-bearing pullulan (CHP)-nanogels in combination with prostaglandin E1 on wound healing, *J. Biomed. Mater. Res. B: Appl. Biomater.*, **91**, pp. 55–60.

144. Bruneel, D. and Schacht, E. (1993). Chemical modification of pullulan: 2. Chloroformate activation, *Polymer*, **34**, pp. 2633–2637.

145. Belbekhouche, S., Ali, G., Dulong, V., Picton, L., and Le Cerf, D. (2011). Synthesis and characterization of thermosensitive and pH-sensitive block copolymers based on polyetheramine and pullulan with different length, *Carbohydr. Polym.*, **86**, pp. 304–312.

146. Kim, D. H. and Rossi, J. J. (2008). RNAi mechanisms and applications, *BioTechniques*, **44**, pp. 613–616.

147. Rekha, M. R. and Sharma, C. P. (2011a). Hemocompatible pullulan-polyethyleneimine conjugates for liver cell gene delivery: In vitro evaluation of cellular uptake, intracellular trafficking and transfection efficiency, *Acta Biomater.*, **7**, pp. 370–379.

148. Juan, A. S., Hlawaty, H., Chaubet, F., Letourneur, D., and Feldman, L. J. (2007). Cationized pullulan 3D matrices as new materials for gene transfer, *J. Biomed. Mater. Res. A*, **82A**, pp. 354–362.

149. Priya, S. S., Rekha, M. R., and Sharma, C. P. (2014). Pullulan-protamine as efficient haemocompatible gene delivery vector: Synthesis and in vitro characterization, *Carbohydr. Polym.*, **102**, pp. 207–215.

150. Wang, J., Dou, B., and Bao, Y. (2014). Efficient targeted pDNA/siRNA delivery with folate low-molecular-weight polyethyleneimine-modified pullulan as non-viral carrier, *Mater. Sci. Eng. C.*, **34**, pp. 98–109.

151. Rekha, M. R. and Sharma, C. P. (2011b). Intracellular trafficking mechanism and cytosolic protein interactions of a nonviral gene delivery vector: Studies based on transferrin conjugated pullulan-PEI, *Curr. Nanosci.*, **7**, pp. 879–885.

152. Zhang, H., Li, F., Yi, J., Gu, C., Fan, L., Qiao, Y., Tao, Y., Cheng, C., and Wu, H. (2011). Folate-decorated maleilated pullulan–doxorubicin conjugate for active tumor-targeted drug delivery, *Eur. J. Pharm. Sci.*, **42**, pp. 517–526.

153. Scomparin, A., Salmaso, S., Bersani, S., Satchi-Fainaro, R., and Caliceti, P. (2011). Novel folated and non-folated pullulan bioconjugates for anticancer drug delivery, *Eur. J. Pharm. Sci.*, **42**, pp. 547–558.

154. Lu, Y. and Low, P. S. (2002). Folate-mediated delivery of macromolecular anticancer therapeutic agents, *Adv. Drug Deliv. Rev.*, **54**, pp. 675–693.

155. Li, F., Zhang, H., Gu, C., Fan, L., Qiao, Y., Tao, Y., Cheng, C., Wu, H., and Yi, J. (2013). Self-assembled nanoparticles from folate-decorated maleilated pullulan doxorubicin conjugate for improved drug delivery to cancer cells, *Polym. Int.*, **62**, pp. 165–171.

156. Hirakura, T., Yasugi, K., Nemoto, T., Sato, M., Shimoboji, T., Aso, Y., Morimoto, N., and Akiyoshi, K. (2010). Hybrid hyaluronan hydrogel

encapsulating nanogel as a protein nanocarrier: New system for sustained delivery of protein with a chaperone-like function, *J. Control. Release*, **142**, pp. 483–489.

157. Sunamoto, J., Sato, T., Hirota, M., Fukushima, K., Hiratani, K., and Hara, K. (1987). A newly developed immunoliposome: An egg phosphatidylcholine liposome coated with pullulan bearing both a cholesterol moiety and an IgMs fragment, *Biochim. Biophys. Acta-Biomembr.*, **898**, pp. 323–330.

158. Taniguchi, I., Akiyoshi, K., Sunamoto, J., Suda, Y., Yamamoto, M., and Ichinose, K. (1999). Cell specificity of macromolecular assembly of cholesteryl and galacto-side groups-conjugated pullulan, *J. Bioact. Compat. Polym.*, **14**, pp. 195–212.

159. Akiyoshi, K., Sasaki, Y., and Sunamoto, J. (1999). Molecular chaperone-like activity of hydrogel nanoparticles of hydrophobized pullulan: Thermal stabilization with refolding of carbonic anhydrase B, *Bioconjugate Chem.*, **10**, pp. 321–326.

160. Bae, B.-C. and Na, K. (2010). Self-quenching polysaccharide-based nanogels of pullulan/folate-photosensitizer conjugates for photodynamic therapy, *Biomaterials*, **31**, pp. 6325–6335.

161. Constantin, M., Fundueanu, G., Bortolotti, F., Cortesi, R., Ascenzi, P., and Menegatti, E. (2007). A novel multicompartimental system based on aminated poly(vinyl alcohol) microspheres/succinoylated pullulan microspheres for oral delivery of anionic drugs, *Int. J. Pharm.*, **330**, pp. 129–137.

162. Dionisio, M., Cordeiro, C., Remunan-Lopez, C., Seijo, B., da Costa, A. M. R., and Grenha, A. (2013). Pullulan-based nanoparticles as carriers for transmucosal protein delivery, *Eur. J. Pharm. Sci.*, **50**, pp. 102–113.

163. Dionisio, M., Braz, L., Corvo, M., Lourenco, J. P., Grenha, A., and da Costa, A. M. R. (2016). Charged pullulan derivatives for the development of nanocarriers by polyelectrolyte complexation, *Int. J. Biol. Macromol.*, **86**, pp. 129–138.

164. Fundueanu, G., Constantin, M., and Ascenzi, P. (2008). Preparation and characterization of pH and temperature-sensitive pullulan microspheres for controlled release of drugs, *Biomaterials*, **29**, pp. 2767–2775.

165. Gu, X.-G., Schmitt, M., Hiasa, A., Nagata, Y., Ikeda, H., Sasaki, Y., Akiyoshi, K., Sunamoto, J., Nakamura, H., Kuribayashi, K., and Shiku, H. (1998). A novel hydrophobized polysaccharide/oncoprotein complex vaccine induces in vitro and in vivo cellular and humoral immune responses

against HER2-expressing murine sarcomas, *Cancer Res.*, **58**, pp. 3385–3390.

166. Hosseinkhani, H., Aoyama, T., Ogawa, O., and Tabata, Y. (2002). Liver targeting of plasmid DNA by pullulan conjugation based on metal coordination, *J. Control. Release*, **83**, pp. 287–302.

167. Ikuta, Y., Katayama, N., Wang, L., Okugawa, T., Takahashi, Y., Schmitt, M., Gu, X., Watanabe, M., Akiyoshi, K., Nakamura, H., Kuribayashi, K., Sunamoto, J., and Shiku, H. (2002). Presentation of a major histocompatibility complex class 1-binding peptide by monocyte-derived dendritic cells incorporating hydrophobized polysaccharide-truncated HER2 protein complex: Implications for a polyvalent immune-cell therapy, *Blood*, **99**, pp. 3717–3724.

168. Kanatani, I., Ikai, T., Okazaki, A., Jo, J.-I., Yamamoto, M., Imamura, M., Kanematsu, A., Yamamoto, S., Ito, N., Ogawa, O., and Tabata, Y. (2006). Efficient gene transfer by pullulan–spermine occurs through both clathrin-and raft/caveolae-dependent mechanisms, *J. Control. Release*, **116**, pp. 75–82.

169. Kim, H. and Na, K. (2010). Evaluation of succinylated pullulan for long-term protein delivery in poly(lactide-co-glycolide) microspheres, *Macromol. Res.*, **18**, pp. 812–819.

170. Lu, D., Wen, X., Liang, J., Gu, Z., Zhang, X., and Fan, Y. (2009). A pH-sensitive nano drug delivery system derived from pullulan/doxorubicin conjugate, *J. Biomed. Mater. Res. B*, **89B**, pp. 177–183.

171. Mocanu, G., Nichifor, M., Picton, L., About-Jaudet, E., and Le Cerf, D. (2014). Preparation and characterization of anionic pullulan thermoassociative nanoparticles for drug delivery, *Carbohydr. Polym.*, **111**, pp. 892–900.

172. Nagane, K., Kitada, M., Wakao, S., Dezawa, M., and Tabata, Y. (2009). Practical induction system for dopamine-producing cells from bone marrow stromal cells using spermine–pullulan-mediated reverse transfection method, *J. Tissue Eng.*, **15**, pp. 1655–1665.

173. Suginoshita, Y., Tabata, Y., Matsumura, T., Toda, Y., Nabeshima, M., Moriyasu, F., Ikada, Y., and Chiba, T. (2002). Liver targeting of human interferon with pullulan based on metal coordination, *J. Control. Release*, **83**, pp. 75–88.

174. Thomsen, L. B., Lichota, J., Kim, K. S., and Moos, T. (2011). Gene delivery by pullulan derivatives in brain capillary endothelial cells for protein secretion, *J. Control. Release*, **151**, pp. 45–50.

175. Wang, B., Chen, K., Yang, R.-D., Yang, F., and Liu, J. (2014). Photoresponsive nanogels synthesized using spiropyrane-modified pullulan as potential drug carriers, *J. Appl. Polym. Sci.*, **131**, 40288.

176. Wang, Y., Liu, Y., Liu, Y., Zhou, W., Wang, H., Wan, G., Sun, D., Zhang, N., and Wang, Y. (2015). A polymeric prodrug of cisplatin based on pullulan for the targeted therapy against hepatocellular carcinoma, *Int. J. Pharm.*, **483**, pp. 89–100.

177. Yim, H., Yang, S.-G., Jeon, Y. S., Park, I. S., Kim, M., Lee, D. H., Bae, Y. H., and Na, K. (2011). The performance of gadolinium diethylene triamine pentaacetate-pullulan hepatocyte-specific T1 contrast agent for MRI, *Biomaterials*, **32**, pp. 5187–5194.

Chapter 9

Transformation of an *Aureobasidium* Pullulan Synthetase Gene into *Saccharomyces cerevisiae*

Anna R. Oller[a] and Carl A. Westby[†]
[a]*University of Central Missouri, School of Natural Sciences, WCM 306 Warrensburg, MO 64093*
oller@ucmo.edu

Aureobasidium pullulans is an environmental fungus known to produce the exopolysaccharide pullulan, which is used in bone repair, drug delivery, hydrogels, nanoparticles, and food applications. The ability to upscale pullulan production is of great interest to many. In this series of experiments, random amplified polymorphic DNA (RAPD) PCR performed on genomic DNA from *A. pullulans* ATCC 42023 (AP1) showed an 1800 bp DNA band, seen upon gel electrophoresis. This DNA band was isolated and transformed into either a pullulan-deficient mutant *A. pullulans* strain or the *Saccharomyces cerevisiae* expression vector pYES2 using either Novozyme or electroporation, and then pullulan elaboration was determined. Induction medium,

[†]Deceased 16 February, 2005

Pullulan: Processing, Properties, and Applications
Edited by Shakeel Ahmed and Aisverya Soundararajan
Copyright © 2021 Jenny Stanford Publishing Pte. Ltd.
ISBN 978-981-4877-17-6 (Hardcover), 978-1-003-05694-2 (eBook)
www.jennystanford.com

reverse transcriptase PCR, and Southern hybridizations all utilized positive and negative controls. Pullulan was quantified using filter weights, pullulanase, and neocuproine. *A. pullulans* (AP1) served as positive control, with an ethyl methyl sulfate mutant (B40203), a non-transformed *S. cerevisiae*, and an antisense *S. cerevisiae* all serving as negative controls. The filter weights of the positive control and the electroporated mutants were all comparable with 42 g/L produced on day 8. The neocuproine estimates were higher than the other two methods, with the positive control at 108 g/L by day 10 and the transformants ranging from 72 to 106 g/L. Novozyme mutants did not restore pullulan synthesis. Statistical analysis showed $p<0.05$ for AP1 and all the electroporated transformed mutants, except the *S. cerevisiae* negative control, which $p = 0.26$. Genetic analysis showed two exons, one 632 bp and the other 218 bp in length, with the 632 bp exon believed to be a synthetase gene. A BLAST search appears to have some homology to two *Aureobasidium melanogenum* pullulan synthetase genes. This research may help the industry to upscale pullulan production so that it could be economically feasible for use in more products.

9.1 Introduction

Aureobasidium pullulans is an environmental fungus known for producing the exopolysaccharide pullulan, which is composed of α-1,6- and α-1,4-linkages that make up maltotriose [6, 13, 31]. The cosmopolitan fungus has been isolated in nature from plant leaves [35], soil and rock [41], and marine environments [33]. The exopolysaccharide pullulan that *A. pullulans* produces is water soluble, tasteless, odorless, impermeable to oxygen [40], nontoxic, can form films [35], is biodegradable, and has many potential industrial applications.

Pullulan's versatile food, agricultural, pharmaceutical, and medical applications range from a biodegradable food coating to prolong the shelf life of berries [26], as a crop application to protect winter wheat from fungal blight [43], a new drug-delivery mechanism for breast cancer [8], and as microbeads to facilitate intervertebral disc repair [21]. Recently, pullulan has also been used in a pH-sensitive hydrogel system with modified-cholesterol-bearing pullulan (CHPOA) [14], in various adhesives [38], as prebiotic nanoparticles in yogurt [22], as antimicrobial enhancers to help combat antibiotic resistance [3], and as a method of preserving bacteriophages [29]. Pullulan could also potentially be used as a natural wood stain [42], or the fungus could be used to remove heavy metals such as cadmium or nickel in environmental remediation situations [10]. Recently, pullulan was combined with the carbohydrate trehalose to produce a thermostable vaccine not requiring refrigeration [28]. Thus, potential applications are vast.

The ability to upscale pullulan production is of great interest to many. Pullulan is expensive to purchase commercially in small quantities, with 1 g currently costing over $70 [49]. Large-scale use of pullulan would become more economically feasible if the production cost could be decreased.

Pullulan elaboration rates by different *A. pullulans* varieties grown on various substrates such as chitosan beads [44], agro-industrial wastes [23], and cellulose [45] have been studied. An et al. [2] found potato starch combined with sucrose-enhanced production, whereas Sheng, Zu, and Tong [36] found that addition of amino acid uracil at specific times enhanced pullulan elaboration. Research continues into determining the optimal conditions for pullulan elaboration by a particular variety or species of *Aureobasidium*.

The *A. pullulans* species are known to have color variants [27], with pigmentation variations seemingly dependent upon whether the culture is grown in sucrose, maltose, or fructose [12]. An inherent problem often encountered with pullulan elaboration and recovery is the production of a black melanin pigment, which then must be removed or bleached during industrial processing. Thus, one scientific goal is to increase pullulan elaboration from the fungus, with a second goal of reducing melanin production. These two accomplishments would greatly increase pullulan production and help reduce overall production costs.

In this study, RAPD PCR was performed on genomic DNA from *A. pullulans* ATCC 42023 (AP1). Four DNA bands were observed upon agarose gel electrophoresis. The DNA bands were isolated, purified, and transformed into either a pullulan-deficient mutant *A. pullulans* strain or the *Saccharomyces cerevisiae* expression vector pYES2 using either Novozyme or electroporation, and pullulan elaboration was then determined using filter weights, the enzyme pullulanase, and a neocuproine assay. Induction medium, reverse transcriptase PCR, and Southern hybridizations all utilized positive and negative controls. *A. pullulans* (AP1) served as a positive control, with an ethyl methyl sulfate mutant (B40203), a non-transformed *S. cerevisiae*, and an antisense *S. cerevisiae* all serving as negative controls.

9.2 Strains and Growth Conditions

9.2.1 *Aureobasidium pullulans*

Aureobasidium pullulans (de Bary) Arnaud has been known since 1891 [9]. *Aureobasidium* comprises three main species: *A. pullulans, A. leucospermi*, and *A. proteae* [31]. With genomic sequencing, *A. pullulans* now consists of five recognized varieties or subspecies: *A. pullulans, A. melanogenum, A. subglaciale, A. thailandense*, and *A. nambiae* [20, 31, 35]. The varieties differ in that *A. pullulans* is the most halotolerant, withstanding up to a 17% salt concentration. *A. pullulans* is often isolated from plants such as apple trees, produces a greenish-black melanin, and grows at 25–28°C. In contrast, *A. melanogenum* is often isolated from aqueous environments, produces a greenish-black melanin, and can grow up to 37°C, whereas *A. subglaciale* is isolated from glacial habitats, produces a greenish-black melanin by 2 weeks post-inoculation onto media, and grows at 4°C. *A. nambiae* is a tropical desert isolate with olive-brown melanin, and *A. thailandense* isolated from leaves and wood in Thailand is known to produce a non-pullulan exopolysaccharide, and produces a reddish-brown melanin [35].

Aureobasidium pullulans (de Bary) Arnaud ATCC® 42023™ (NRRL Y-12996) was purchased from the American Type Culture Collection (Manassas, VA, USA) and was used as the wild-type (AP1) pullulan-producing, uracil-positive (*URA*3+) strain. This strain is also known to produce exopolymers, and the enzymes pectinase and polygalacturonase, among others [30]. *A. pullulans* is known to produce many enzymes such as xylanase [27], pectinase, amylase, and protease [9]. According to Gostinčar et al. [20], the *A. pullulans* in their study was shown to possess a Type 1 polyketide synthase, a non-ribosomal peptide synthase, chitinases, proteases, glycoside hydrolases, polysaccharide lyases, many peptidases, tannases, and sugar transporters.

Orotidine-5-monophosphate (OMP) is involved in pyrimidine synthesis [4, 25, 32]. Decarboxylation of orotidine-5-monophosphate to uridine monophosphate is often used as a selection marker for the *URA*3 gene [3, 7, 18]. The orotidine-5-monophosphate (OMP) enzyme is thought to be a single function enzyme, and mutations can lead to autotrophy of uracil [32]. Thus, OMP decarboxylase enzyme activity can be quantitated, and 5-fluoro-orotic acid (5-FOA) can be added to agar to select for mutant revertants [7, 36, 46].

All *A. pullulans* strains were grown aerobically at 28°C in a rotary shaker at 200 rpm, unless otherwise noted, and are shown in Table 9.1. An ethyl methyl sulfate treatment followed by gamma irradiation yielded a pullulan-deficient mutant of *A. pullulans* (B40203) characterized by Li [30] and served as a negative control. Yeast potato dextrose (YPD) consisting of 1% yeast extract, 2% peptone, and 2% α-D-glucose was used for experiments and stock cultures. Modified Ueda (MU) was used for polysaccharide elaboration and consisted of 0.5 g K_2HPO_4, 0.1 g NaCl, 0.02 g $MgSO_4 \cdot 7H_2O$, 0.06 g $(NH_4)_2SO_4$, and 10% (*w*/*v*) sucrose, as described by Oller [34]. For mutant selection, synthetic medium with (SMU) and without (SM) uracil consisting of 1 g K_2HPO_4, 5 g $MgSO_4 \cdot 7H_2O$, 0.05 g KCl, 0.01 g $FeSO_4 \cdot 7H_2O$, 3 g NH_4Cl, and 1% sucrose (*w*/*v*) per liter, as well as 5-fluoro-orotic acid (5-FOA) (0.1 g/15 ml) were utilized. Plates were made using 2% agar (Difco Laboratories Inc., Detroit, MI, USA).

9.2.2 Gamma-Irradiated Mutants and OMP Decarboxylase

Aureobasidium pullulans ATCC 42023 (AP1) was grown overnight in YPD, with shaking at 200 rpm. The 5 ml of starter culture was added to a new flask containing 40 ml of fresh YPD. The pre-exposure optical density (OD) at 525 nm was determined, and 3 ml of fresh culture was aseptically added to sterile 10 mm glass test tubes. Tubes were exposed to the gamma irradiation (cobalt) source for 0, 20, 30, or 40 min. Serial dilutions were made to plot kill curves, and 100 μl from each tube was spread onto YPD, SMU, SM, and YPD containing 5-FOA. Plates were checked daily and incubated for up to 7 days. Colonies were subcultured onto fresh YPD, SMU, SM, and Ueda plates to look for uracil-negative (*URA*3−) and pullulan-positive strains.

To ensure mutants were truly both pullulan and uracil deficient, OMP decarboxylase and protein curves were performed multiple times. Initial results of the OMP decarboxylase activity revealed a level of 0 for the B40203 mutant, whereas AP1 was 20 g/L when plotted against the standards in the protein curve. Overall, the enzymatic activity was calculated to be 0.12 μmol/min/mg.

Table 9.1 Strains and plasmids used in experiments

Strain/Plasmid	Genotype/Description	Source
E. coli		
INVαF′	F′ (*end*A1 *rec*A1 *hsd*R17 ($r_k^{-}m_k^{+}$) *sup*E44 *thi*-1 *gyr*A96 *rel*A1 φ80*lac*ZΔM15 Δ(*lac*ZYA-*arg*F)U169λ⁻	Invitrogen
TOP10F′	F′{*lac*I^{q}Tn*10*(Tet^R)}*mcr*A Δ(*mrr-hsd*RMS-*mcr*BC)φ80*lac*Z ΔM15 Δ*lac*X74 *deo*R *rec*A1 *ara*D139 Δ(*ara-leu)*7697 *gal*U *gal*K *rps*L *end*A1 *nup*G	Invitrogen
S. cerevisiae		
INVSc1	(MATα *his* 3Δ1/*leu*2 *trp*1-289 *ura*3-52)	Invitrogen
Strains 1-5	PCR2.1 and pYES2 mutants with restored pullulan production	Laboratory stock
A. pullulans		
ATCC 42023 (AP1)	pullulan-producing strain, *URA*+	Laboratory stock
B42023	EMS and gamma irradiated 42023 mutant; pullulan and melanin deficient, *URA*-	Laboratory stock
Trans. Strain #6	pullulan-producing strain of B40203 protoplasts, *URA*+	Laboratory stock
Plasmids		
pCR2.1-TOPO	vector for cloning PCR products	Invitrogen
pYES2	yeast expression vector containing *GAL* promoter	Invitrogen

9.3 DNA Amplification and Gel Electrophoresis

A Puregene genomic DNA isolation kit for Gram-positive bacteria and yeast (Qiagen, Hilden, Germany) was used to isolate genomic DNA from both pullulan-deficient gamma-irradiated mutants and wild-type strains. The lytic enzyme incubation time was extended to 1 h to ensure adequate DNA quantity recovery due to surface polysaccharides. A 10-mer primer O-AB04 (Operon Technologies Inc., Alameda, CA, USA) with the sequence of 5′-GGCACGCGTT-3′ was utilized in the polymerase chain reaction (PCR). A 25 μl PCR reaction contained 35 ng of genomic DNA, 4.5 μM O-AB04 primer, 4.5 mM dNTP mix (Promega, Madison WI, USA), 10 × buffer containing 15 mM $MgCl_2$, and 0.8 U Taq Polymerase (Sigma-Aldrich, St. Louis, MO, USA). A thermocycler (Perkin Elmer, Boston, MA, USA) was programmed for 45 cycles of: 94°C for 30 s, 30°C for 1 min, and 72°C for 2 min with a final extension cycle of 72°C for 10 min.

A 1% tris-acetate-EDTA (TAE) agarose gel (Bio-Rad, Hercules, CA, USA) containing 50 μl of ethidium bromide (1 μl/ml) was submersed in 1× TAE buffer. A 1 kb ladder served as a standard for each run. The gel was electrophoresed at 70 V for approximately 1.5 h and then visualized under a UV transilluminator. Four DNA bands (Fig. 9.1) were excised, purified using a filter spin column (Ambion), re-amplified, and electrophoresed.

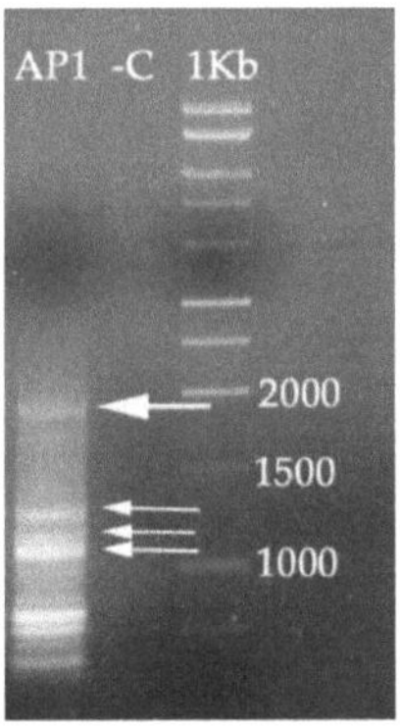

Figure 9.1 A 1% TAE agarose gel depicting four bands seen in *Aureobasidium pullulans* ATCC 42023 (AP1) using the RAPD primer O-AB04. The large arrow at the top shows the 1800 bp band matching a pullulan synthetase gene of *A. melanogenum*. The other three bands at 1300 bp, 1200 bp, and 1100 bp were used in experiments but did not restore pullulan secretion.

9.3.1 Plasmid Vectors

Once DNA bands were excised from the agarose gel and purified, the plasmid pCR2.1-TOPO (Invitrogen Co., San Diego. CA) was used for transforming the PCR products into competent *Escherichia coli* (INVαF) cells (Invitrogen Co.), which contained blue/white colony selection. The yeast expression vector pYES2 (Invitrogen Co.) was used for introducing DNA segments into *E. coli*, which were subsequently transformed into *S. cerevisiae* (strain INVScl) (Invitrogen Co.) to determine gene expression.

The four DNA bands (Fig. 9.1) (1800 bp, 1300 bp, 1200 bp, and 1100 bp) were each ligated into the plasmid pCR2.1 from the TOPO-TA cloning kit and transformed into *E. coli* cells in triplicate, utilizing blue/white colony selection (Fig. 9.2). Bacterial cultures were grown at 37°C in a rotary shaker at 200 rpm, to log phase in 50 ml of Luria Bertani (LB) broth (BIO101, La Jolla, CA, USA) supplemented with 5-bromo-4-chloro-3-indolyl-beta-D-galacto-pyranoside (XGAL), isopropyl-beta-D-thiogalacto-pyranoside (IPTG), and either 50 μg/ml of ampicillin (pYES2) or kanamycin (pCR2.1-TOPO) per Invitrogen directions.

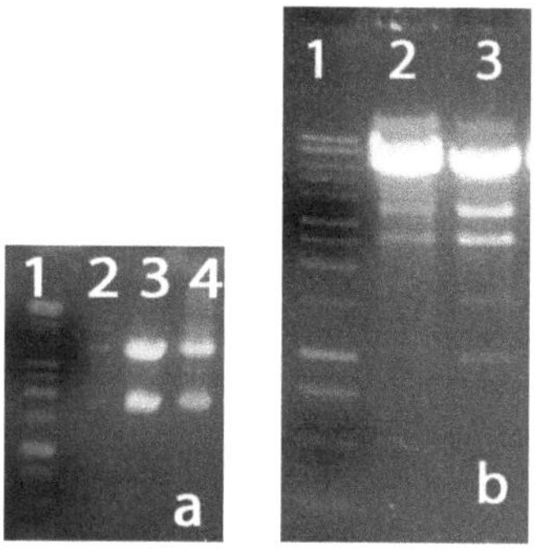

Figure 9.2 A 1% TAE agarose gel with ethidium bromide depicting (a) the 1800 bp band ligated into the *E. coli* pCR2.1 vector using *EcoR*1 and (b) the sense ligation into the pYES2. Lane 1 on both is a 1 kb ladder.

Bacterial cells were either digested with the enzyme *EcoR1* or lysed by heating and used directly in a PCR and the insert was confirmed by visualization on an agarose gel. Plasmid DNA was isolated and purified from the cloned *E. coli* cells using the Rapid Pure

Miniprep for Automated Fluorescence Sequencing (RPM-AFS) Kit manufacturer's directions (BIO 101). Purified plasmid DNA was then quantified on a Beckman DU-64 spectrophotometer. All experiments were performed at minimum in duplicate.

9.3.2 Novozyme DNA Transformation

Isolated plasmid DNA described above was used in transformation experiments carried out in duplicate, as described by Thornewell et al. [39] with the following modifications. The B40203 negative control culture was incubated at 28°C for 16 h with shaking at 200 rpm. Novozyme 234 from *Trichoderma harzanium* (Sigma-Aldrich) was used to produce protoplasts. Both linear (cut with *Eco*RI) and circular plasmid DNA (100–125 μl) was added to 50 ml centrifuge tubes and incubated for 10 min before 5 ml of a 50% polyethylene glycol (PEG) solution was added. The transformation suspension was incubated, with shaking, for 2 h at room temperature and then spread onto Trypticase soy agar (TSA), YPD, SM, SMU, and MU plates. Plates were checked daily and incubated for up to 7 days. The five resulting colonies were then subcultured onto MU plates to visualize pullulan production. Transformants were subcultured a second time to ensure DNA stability and then were subcultured further onto YPD, Ueda, and cryopreserved with glycerol at −20°C. Only one colony was successfully recovered, but it did not appear to restore pullulan production. Previous studies have shown various strains of *A. pullulans* to have genetic instability and must be re-cloned into another strain [33]. This is precisely the approach taken in this study.

9.3.3 B40203 Protoplast Transformation

Cullen et al. [15] previously used protoplasts to transform a hygromycin-resistance gene into *A. pullulans*, so protoplasts were utilized in this study to try to transform the pullulan-deficient B40203 strain with the PCR bands isolated from agarose gels. The method of Finkelman [19] was performed in duplicate to create yeast protoplasts and introduce each PCR band of DNA into the genome of the pullulan-deficient strain, B40203. Shiny, viscous, mucoid appearing colonies were transferred to a Ueda plate, but no pullulan production was seen after the transfer, indicating the DNA appears unstable in the genome. Transformants were subcultured a second time to ensure DNA stability and then were subcultured further onto YPD, Ueda, and cryopreserved with sterile glycerol at −20°C. One colony transformed with the 1800 bp DNA appeared to restore both melanin and pullulan production and was designated Transformant Strain #6.

9.4 Photography

Fungal photographs of colonies grown on agar media were obtained using a Microleaf digital camera (Leaf America, Elmsford, New York). On post-inoculation days 2, 4, 6, 8, and 10 of pullulan quantification, culture samples were added to wet mount slides and viewed microscopically on an Olympus Provis AX70 upright microscope at 400 × total magnification.

Photos of the wild type (AP1), the pullulan-deficient mutant B40203, and Transformant Strain #6, a B40203 restored mutant, are shown in Figs. 9.3, 9.4, and 9.5, respectively.

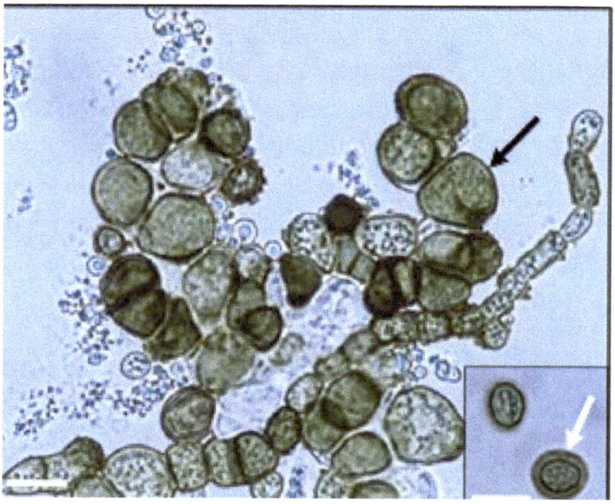

Figure 9.3 The wild-type AP1 showing pullulan being elaborated on day 6. Various cell types are seen. Scale bars are 10 μm.

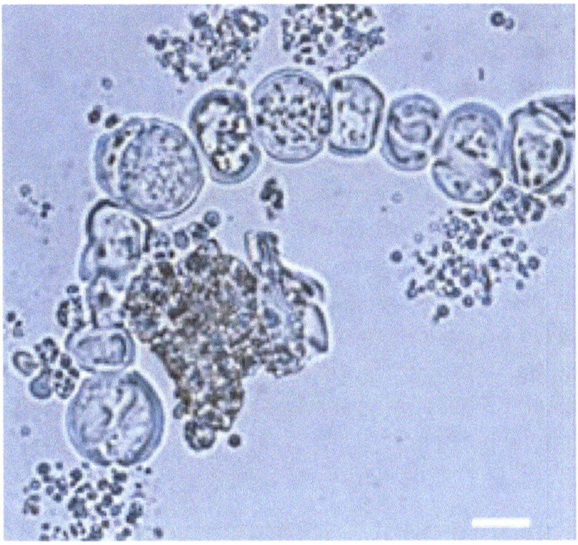

Figure 9.4 Gamma-irradiated pullulan-deficient strain B40203 showing no melanin or pullulan being elaborated on day 8. Scale bars are 10 μm.

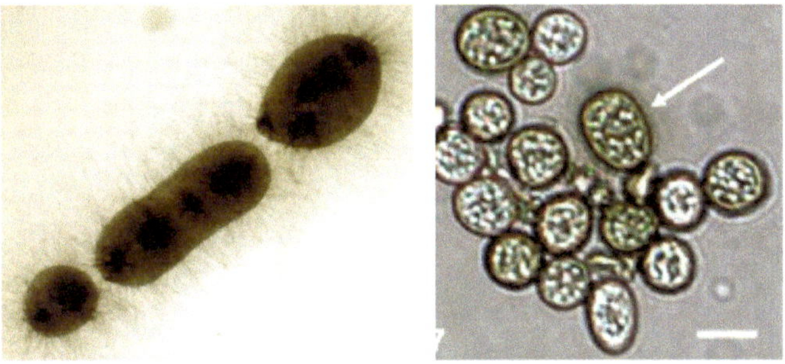

(a) (b)

Figure 9.5 Transformant Strain #6 showing restored pullulan synthesis (a) on an MU plate and (b) microscopically at 400×. An 1800 bp band was transformed into a pullulan-deficient strain, B40203. The pullulan did not become colored until days four to six and was less viscous than the wild-type pullulan. Scale bars are 10 μm.

9.5 Southern Hybridization

Agarose gels containing the DNA bands of interest were denatured, neutralized, and transferred to a nitrocellulose membrane according to the photogene nucleic acid detection system. The dNTPs and hexanucleotides were then mixed with 50 μCi of ^{32}P dATP (Sigma-Aldrich) and 1 μl of Klenow and incubated for 60 min at 37°C. The reaction was stopped using EDTA and heated at 65°C for 10 min.

A Sephadex G-50 column was set up with 1× TENS used as elution buffer, and a scintillation counter determined which tube contained the highest radioactivity to use as the probe. The membrane was placed in a 42°C oven in church buffer for 2 h. Autoradiography film was then placed directly against the membrane in a −20°C freezer and exposed for various times of exposure time and then developed. The PCR bands showing the best likelihood at being a gene of interest, and fungal strains serving as controls, were then utilized in subsequent experiments.

9.6 Recombinant *Saccharomyces cerevisiae* Strains

Plasmid pCR2.1-TOPO was used for transforming the four *A. pullulans* PCR products into competent *E. coli* INVαF-TOPO cells, and white colonies were selected. Plasmid pYES2 introduced DNA segments into *E. coli*, which were subsequently transformed into *S. cerevisiae* INVSc1 cells to determine gene expression. RNA was isolated from *S. cerevisiae* transformants using RNAWIZ solution (Ambion, Inc., Austin, TX). A Rapid Pure Miniprep for Automated Fluorescence Sequencing (RPM-AFS) kit (BIO 101) was used for plasmid DNA isolation. Cell walls of *S. cerevisiae* were lysed using 100 mg/ml of lytic enzyme (Sigma-Aldrich, St. Louis, Missouri) and RNA isolation was performed using RNAWIZ solution (Ambion, Inc, Austin, Texas).

The modified DNA of the 1800 bp PCR band was then ligated into the pYES2 vector by incubating the following 10 μl reaction for 14 h at 14°C: 4 μl of Insert DNA, 1 μl of 10× Ligation buffer, 1 μl of T4 DNA Ligase (4.0 Weiss units), 0.25 μl of pYES2 vector (25 ng/μl), and 3.75 μl of sterile double distilled (dd) H_2O. The vector pYES2 was then transformed into competent *E. coli* cells as described by the Invitrogen manual. *E. coli* plasmid DNA was isolated using the RPM-ASF kit.

Several restriction enzymes were then used to create appropriate overhangs for both forward (sense) and backward (antisense) ligation of the 1800 bp PCR DNA into pYES2. The vector was cut with the enzymes *Not*I (15 U/μl) and *Bam*HI (10 U/μl) (Gibco Life Technologies, Grand Island, New York) to ligate the sense gene. The antisense ligation served as a negative control since gene transcription should not occur in this direction. However, *Not*I is absent in the multiple cloning site (MCS), so a 5′-phosphorylated linker (5′-GGGCCGGTA-3′) purchased from Integrated DNA Technologies (Coralville, Iowa) was used to convert the *Not*I site to a *Kpn* site, which is present in the MCS.

Ligation of the linker DNA was carried out using T4 ligase and buffer according to directions provided by Sigma-Aldrich and incubated at 14°C for 16 h. The antisense sequence was ligated into pYES2 using *Xba*I and *Hin*dIII (both supplied in a concentration of 10 U/μl), to create overhangs, which are both present in the MCS. The linker DNA was ligated to the presumptive 1800 bp gene in a 20 μl reaction (10 μl DNA, 2 μl 10× T4 DNA ligase buffer, 5 μl T4 DNA ligase (Sigma-Aldrich), 3 μl linker DNA) and incubated at 14°C for 16 h.

Next, the enzyme *Kpn* (10 U/μl) was used to cut the DNA to provide an appropriate overhang. The modified 1800 bp DNA was then ligated into pYES2 after incubation for 14 h at 14°C using the following 10 μl reaction: 4 μl Insert DNA (5 μg), 1 μl 10× ligation buffer, 1 μl T4 DNA ligase (4 Weiss Units), 0.25 μl pYES2 vector (25 ng/μl), and 3.75 μl sterile ddH_2O. The vector pYES2 was transformed into TOP10F′ *E. coli* cells, and then subsequently electroporated into *S. cerevisiae* cells. Restriction enzyme digestion and visualization on agarose gels confirmed the orientation of the DNA sequence, as shown in Fig. 9.6.

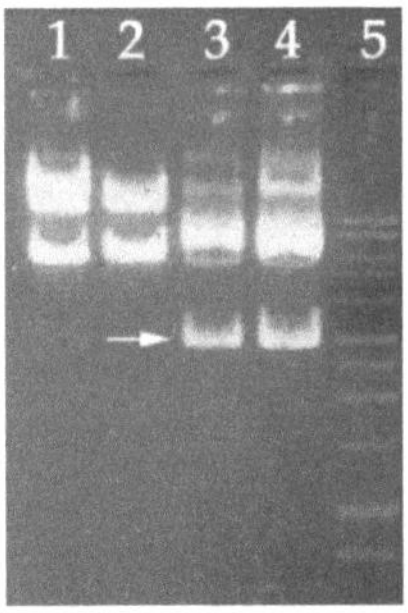

Figure 9.6 A 1% TAE agarose gel with ethidium bromide depicting ligated bands into the *E. coli* pYES2 vector. Lane 1 is the antisense ligation (faint band). Lane 2 is a negative control. Lanes 3 and 4 are sense ligations, and Lane 5 is a 1 kb ladder. The arrow shows the band of interest.

Saccharomyces cerevisiae was grown in 15 ml of SC Minimal Media (Invitrogen, San Diego, CA, USA) containing uracil at 30°C to identify *S. cerevisiae* transformants. Induction media, with galactose and raffinose added, promoted gene expression in the transformants. Cultures were cryopreserved in 20% (v/v) of sterile glycerol at −80°C.

9.7 Electroporation of *Saccharomyces cerevisiae*

Because the *E. coli* plasmid could not express pullulan, the plasmid then needed to be added to *S. cerevisiae* so expression could occur and be measured. *S. cerevisiae* was grown overnight in 50 ml of YPD broth at 30°C at 200 rpm. Competence was induced by the method of Amberg [1] by employing cold water, 1 M sorbitol, and 25 mM dithiothreitol (DTT). Isolated pYES2 DNA (20 μg) containing either sense or antisense (control) insert was added to the mixture. A negative control using ddH_2O was also performed.

Electroporation in 1 mm gap cuvettes was carried out at 50 ms/−150 v, 5 ms/−150 v, and 10 ms/−588 v. The electroporated cells were then mixed with 1 ml of YPD/1 M sorbitol and incubated at 25°C for 1 h. Electroporated cells (333–500 μl) of the mixture were plated as a lawn of growth onto both SM and SMU medium and allowed to grow for up to 5 days. Colonies were then subcultured onto fresh SM, SMU, and MU to look for growth selectivity, pigmentation, and pullulan production. *S. cerevisiae* was then cultured in induction broth, with samples being taken every two h for up to 14 h. RNA was then isolated from each sample as previously mentioned. Transformants were cultured in both modified Ueda and induction medium containing the sugars galactose and raffinose. Five sense and two antisense transformants were cultured in induction broth, with 2.5 ml removed every two h for 14 h for RNA isolation. RNA was then isolated as previously mentioned.

9.8 Reverse Transcriptase PCR

To detect sense and antisense expression, reverse transcriptase (RT) was employed using a Molony–Murine Leukemia Virus and other reagents supplied with the Retroscript First Strand Synthesis Kit for RT-PCR (Ambion). For RAPD-PCR of the cDNA, Supertaq *Taq* Polymerase (Ambion) optimized for the Retroscript kit was utilized, and the annealing temperature for the O-AB04 primer was used. Appropriate negative controls (supplied with kit, *S. cerevisiae* not transformed, *S. cerevisiae* antisense, water as a blank, etc.), were included in each process. Two-step RT-PCR was performed according to the Ambion manual, in which 2 μg of template RNA was used. The following parameters were programmed into the thermocycler: 1 cycle of 95°C for 5 min, 30 cycles of 94°C for 30 s, 30°C for 30 s, and 72°C for 1 min, and a final cycle of 72°C for 5 min.

9.9 Pullulan Quantification

Once transformants were isolated and determined to appear to produce pullulan, several methods were employed to quantify the pullulan being secreted. For the wild type (AP1), negative control (B40203), five *S. cerevisiae* sense, and two *S. cerevisiae* antisense transformants, pullulan elaboration was quantified using filter weights and pullulanase as described by Dygert [17], and neocuproine using the method described by Israilides et al. [24] with the following modifications.

9.9.1 Filter Weights

First, filter weights of the samples were determined for each transformant, the positive control (AP1) and negative control (B40203). All percentages are on a weight/volume (w/v) basis unless otherwise noted. Every 2 days post-inoculation (days 2, 4, 6, 8, and 10), 5 ml of sample was aseptically removed from shaking flasks and was mixed with 5 ml of absolute ethanol. The precipitated pullulan was vacuum filtered onto pre-weighed Whatman #3 filter paper (50 mm circles) (Sigma-Aldrich) and dried to a constant weight in an oven. The pre- and post-filter weights were then subtracted to determine the amount of pullulan elaborated. All filter weights were performed in duplicate.

Second, 1 ml of sample was mixed with 2 ml of absolute ethanol, which was also vacuum filtered onto the Whatman #3 filter paper.

The filter paper was then placed in a beaker to which 2 ml of 0.2 M phosphate buffer was added and mixed to obtain a uniform mixture. The reagents were added to disposable cuvettes and run on the spectrophotometer as described by Israilides et al. [24]. One run contained no pullulanase (water was substituted) to obtain a background reactivity level. The second run contained pullulanase from *K. pneumoniae* (Sigma-Aldrich) (0.3 U/ml), and a negative control was included on each run. In order to determine a standard curve, known concentrations of pullulan (Hayabashira Co, Japan) (10 g/L, 5 g/L, 2.5 g/L, 1.25 g/L, and 0.62 g/L) were dissolved in 0.2 M phosphate buffer and run twice on the spectrophotometer as previously described. Graphs and reactivity results were calculated by the spectrophotometer.

9.9.2 Neocuproine and Pullulanase

Pullulan quantities utilizing a copper and neocuproine reaction were also determined [17]. Pullulanase was prepared using 0.3 U/ml to obtain glucose molecules. Copper (II) reacts with glucose, and upon boiling, the neocuproine serves as a copper chelator and a colored product is formed. Pullulan standards using the purified pullulan mentioned above were prepared (5, 10, 15, 20, 25, 50, 75, 100, and 125 μg/ml). Spectrophotometer readings were taken every 15 min for 1 h. All runs were performed in duplicate every 2 days as previously mentioned and negative controls utilized sterile water.

9.10 Results

9.10.1 Pullulanase Assay

Five 6 mm mucoid colonies containing the 1800 bp band were seen. Pullulan was seen on Ueda plates, and many fungal stages were seen. The melanin appeared light yellow in the center surrounded by dark green. One colony appeared yellow, which subsequently turned a light to dark green color as the colony aged. Several small mucoid colonies were observed after transformation with the 1300 bp and 1200 bp bands. However, upon subculturing to a fresh modified Ueda plate, no pullulan was seen.

The five mucoid colonies were subjected to growth and the pullulanase assay. Since the colonies did not vary much in their production, they were averaged and placed in Fig. 9.7 for comparison to AP1 and B40203. The colonies did not produce as much pullulan as the positive control. The graph in Fig. 9.7 shows the positive and negative control, averaged *S. cerevisiae* transformants, Transformant Strain #6, and the average Novozyme transformants. The Transformant Strain #6 showed slower production, but more pullulan was elaborated by day 10.

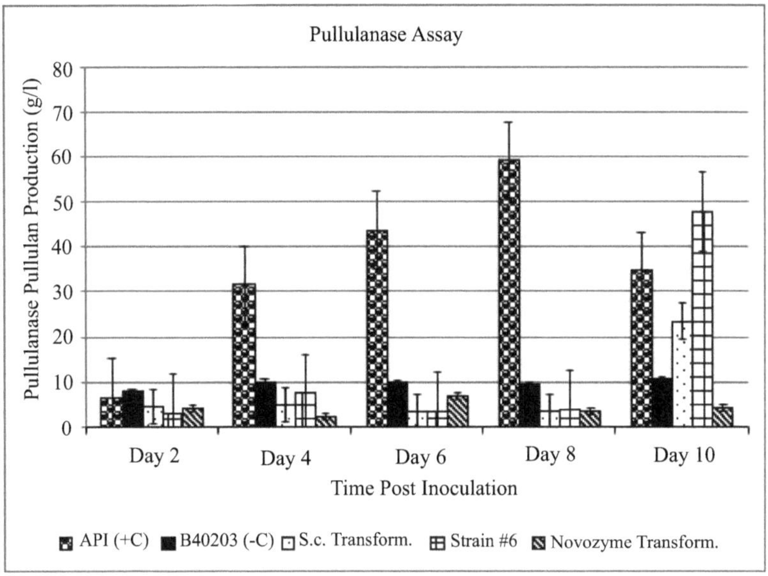

Figure 9.7 Pullulanase enzyme assay showing pullulan quantities from *A. pullulans* (AP1), a pullulan-deficient ethyl methyl sulfate mutant (B40203), *Saccharomyces cerevisiae* transformants, and Novozyme transformants. Standard error bars are shown.

9.10.2 Electroporation

The results of the electroporated transformed *S. cerevisiae* cells appeared more successful than the Novozyme transformants. At 50 ms/−150 v, two transformants were obtained; 11 transformants were obtained at 10 ms/−588 v, and 82 transformants were obtained at 5 ms/−150 v. The negative control exhibited approximately the same number of viable cells, which all appeared as normal, off-white *S. cerevisiae* colonies. All transformants exhibited an exopolysaccharide on the surface of the colonies. After several days, the pullulan did not dry up on the agar plate like the wild type—it could be found on the interior of the dried colony after the surface was punctured with a loop (Fig. 9.8).

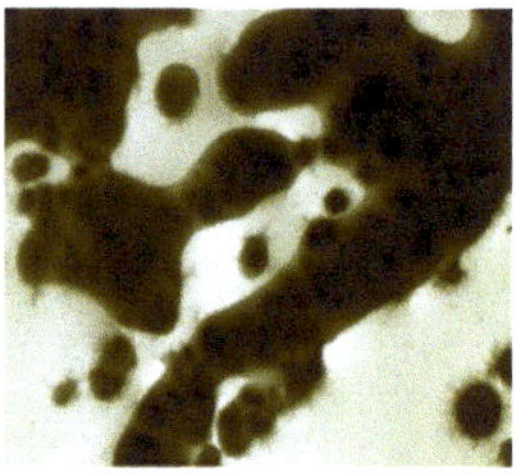

Figure 9.8 *Saccharomyces cerevisiae* transformant, pigmented and producing pullulan on MU agar.

9.10.3 Filter Weights

Filter weights were obtained for B40203 transformants, as well as the *S. cerevisiae* pYES2 transformants, including the negative control. The results showing pullulan filter weights over the 10-day sampling period are shown in Fig. 9.9. *S. cerevisiae* cells transformed with the antisense DNA, yielded yeast cells with little to no pullulan produced, and served as a negative control. No pigmentation or pullulan was seen, even at day 10. All pYES2 strains appeared to increase their pullulan output every 2 days, up until days 6 or 8, depending on the strain. Microscopy results showed that all four *S. cerevisiae* transformants and B40203 Transformant Strain #6 began producing pullulan on day 2, whereas the wild type did not begin producing visual pullulan microscopically until day 4.

Figure 9.10 compares pullulan levels between the strains, measured via filter weights on each sample day. On day 4, all mutants produced more pullulan than the wild type. B40203 mutant #6 did not increase production between days 4 and 6, but significantly increased its output to outpace AP1 by day 8. Further, both B40203 and *S. cerevisiae* negative controls clearly demonstrated a lack of pullulan production.

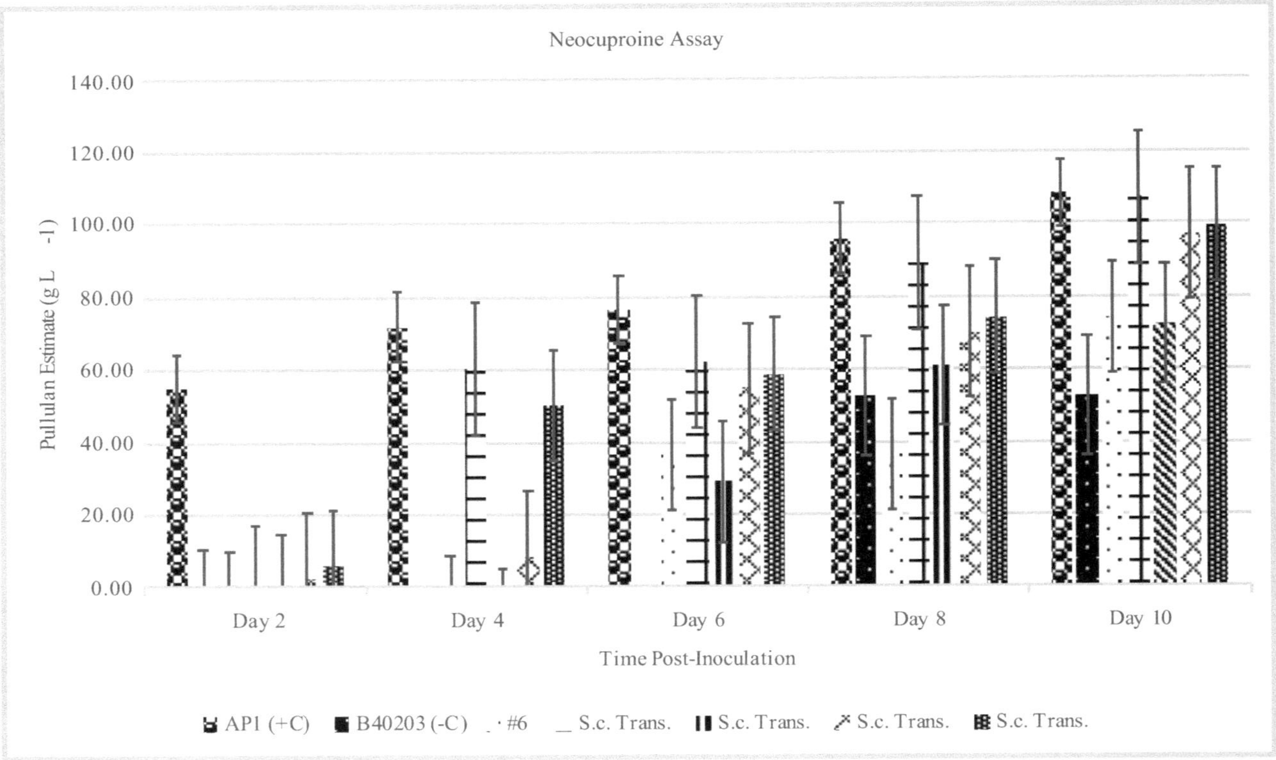

Figure 9.9 Filter weight pullulan production between controls and transformants. Samples were placed in absolute ethanol, filtered onto pre-weighed filters, dried, and subjected to the pullulanase enzyme. Bars represent standard error.

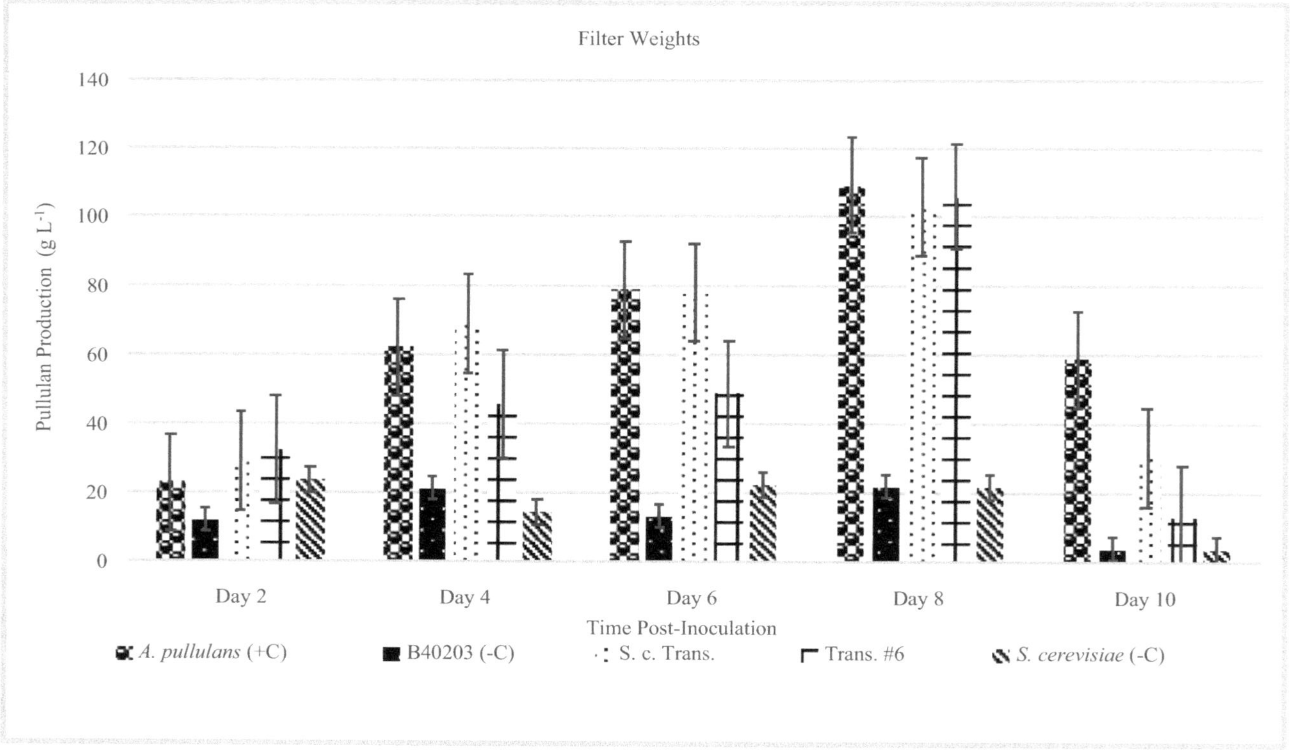

Figure 9.10 Filter weight pullulan production between controls and transformants. Samples were placed in absolute ethanol, filtered onto pre-weighed filters, dried, and subjected to the pullulanase enzyme. Bars represent standard error.

9.10.4 Reverse Transcriptase PCR

Reverse transcriptase PCR was performed on *S. cerevisiae* cells to determine if gene expression was induced and the optimum expression time. RT-PCR results demonstrated that the pullulan gene was being expressed, with the optimum expression time at 10 h. Upon excising the band from an agarose gel and reamplifying it, a single band was detected in the sense transformant and nothing was observed in the antisense transformant (Fig. 9.6).

9.10.5 Microscopy

Microscopy revealed pullulan characteristics. Each organism that was analyzed for filter weights was also viewed microscopically on each sample day (data not shown). AP1, B40203, and B40203 mutants showed polymorphic stages of yeast, blastospores, and conidiospores. Melanin and pullulan appeared at day 4, and the culture continued through day 8, where the cells began to form large clumps as cells stuck together. The B40203 Transformant Strain #6 showed the same stages as AP1, but overall more yeast cells were seen. All *S. cerevisiae* transformants contained the yeast cell stage. The *S. cerevisiae* negative control contained un-pigmented yeast cells.

9.10.6 Statistical Analysis

Filter weight statistics were performed using ANOVA. The AP1 once again served as a positive control, and both B40203 and *S. cerevisiae* pYES (antisense) served as negative controls. When all strains were compared to B40203, the p-value was <0.05 for AP1 and all the mutants, except for the *S. cerevisiae* negative control, which was 0.26. When the p-values were compared to AP1, only the two negative controls had values $p<0.05$. Thus, the level of pullulan produced by each mutant was significant. Standard error bars are also shown in Figs. 9.7, 9.9, and 9.10.

Two glucose enzymatic reactions were employed to determine the quantity of pullulan secreted by *A. pullulans* ATCC 42023 and B40203 to ensure B40203 was a true negative control. Pullulanase determination was calculated in g/L based on the standard curve.

Since ATCC 42023 quantities were much higher than the 10 g/L standard, the curve was extrapolated using the Forecast command in Excel. Since samples were diluted by a factor of 15, this number was programmed into the spectrophotometer and final reaction rates were calculated. All negative control background reactivities were subtracted from the pullulanase-treated calculations. For the 5 and 10 g/L standards and ATCC 42023, the enzyme activity peaked between 18 and 19 min. Pullulanase comparisons between ATCC 42023 and B40203 can be seen in Fig. 9.9. The *A. pullulans* ATCC 42023 peaked at 94 g/L, whereas B40203 peaked at 16 g/L. B40203 levels remained fairly constant with SD ± 2.6 g/L, but ATCC 42023 fluctuated with SD ± 30 g/L.

The neocuproine and copper reaction did not appear as sensitive to low quantities of pullulan since the B40203 peaked at 52 g/L compared to the 10 g/L in the pullulanase assay. The wild type peaked at 108 g/L, compared to the 60–66 g/L detected by the pullulanase assay. The SD was ±33 for ATCC 42023 and ±14 for B40203.

The neocuproine assay averages ($n = 4$) for each were significant at $p = 0.0001$, even when the transformants were lumped together, $p = 0.02$.

9.10.7 DNA Sequencing

Each DNA band was sequenced, and the 1800 bp band was primer walked (Iowa State University Sequencing Facility). Gene Jockey determined that the 1200 bp and 1100 bp sequences were identical, but the 1300 bp sequence was different from the other two. No homology was found between the 1300 bp or 1200 bp bands and the 1800 bp band. Gene Jockey analysis indicated an ORF and Exon 1 at 1018 bp through 1650 bp, and Testcode Algorithm predicted a 95% probability of this region being a coding region. Further analysis with GRAIL indicated two exons: a 3′UT region and a poly-A signal for one gene (1018-1650). Genetic analysis showed two exons: one 632 bp and the other 218 bp in length, with the 632 bp exon believed to be a synthetase gene. A BLAST search appears to have some homology to two *Aureobasidium melanogenum* pullulan synthetase genes. The pullulan synthetase gene sequence was deposited into Genbank under Nucleotide Accession Number AF470619.

9.11 Discussion

9.11.1 Pullulan Secretion

Pullulan elaboration and cell structures were viewed microscopically. The wild-type fungus exhibited a smooth cell membrane that appears to have pullulan elaborated within the cell, which then appeared to be transported to the outside of the cell, agreeing with Campbell et al. [11]. The outer cell membrane began to show a light green fuzzy appearance on day 4, which turned into a black line by day 10. In contrast, B40203 shows a rough cell membrane with no pigment within the cell. PCR studies and Southern hybridization studies of other pullulan-deficient mutants (obtained by Li [30]) performed in our laboratory showed that the gene is still present in some strains. Thus, a disrupted cell membrane possibly contributes to the lack of pullulan production (possibly by disturbance of docking proteins and other signaling mechanisms) since all mutants with no pullulan elaboration contain a disrupted cell wall.

Further, the presence of a pullulan-transporter mechanism [13] is strengthened by the finding that dry *S. cerevisiae* transformant cells still contained pullulan within the cell. When the electroporated sense cells were touched with an inoculating loop for transferring, the liquid escaped from inside of the cell and was so viscous that it caused problems during transfer.

Saccharomyces cerevisiae transformants appeared to grow solely in the yeast phase. Since no other stage resembling AP1 or B40203 was seen, the pullulan expression gene did not seem responsible for inducing other cell stages such as blastospores or conidiospores.

9.11.2 Genetic Analysis

Software analysis indicated the presence of two exons: one 632 bp in length and the second one 218 bp in length. Based on the exon lengths, visual colony characteristics, and the enzyme results, it appears that the 632 bp exon is a pullulan synthetase gene or a polysaccharide-producing gene based on our results. In addition, the 218 bp exon may be responsible for producing a melanin or pigment. Visual colony characteristics revealed that every B40203 (pullulan-deficient cells) transformant with the 1800 bp band DNA resulted in various amounts of melanin being produced. Some colonies appeared as dark as the wild type (AP1), whereas others contained lighter hues of green, and even some yellows. Cells transformed with the 1300 bp and 1200 bp bands showed a range from completely white cells to black cells. Very little pullulan was produced by these cells, and since subculturing resulted in no pullulan production, it appears these genes were unstable, or unregulated, when randomly inserted into the genome. In addition, every *S. cerevisiae* transformant also contained a pigment to varying degrees, although most cells were the color of AP1. Since *S. cerevisiae* negative controls were used throughout and none developed color characteristics, it appears that a pigment or melanin gene and synthetase gene appear closely coregulated or integrated.

9.11.3 Transformants

Ma et al. [33] and Li et al. [31] appear to have identified a pullulan synthetase gene from a different *A. pullulans* strain, HN6.2, than the current study. However, since a presumptive pullulan synthetase gene has been identified in Genbank, this could be very beneficial to the industry. By creating pullulan overproducers in *S. cerevisiae* to the industry, easier culturing methods could be employed, making pullulan production for the industry cheaper and easier. Before the industry can use *S. cerevisiae* pullulan transformants in fermented beverages as a value-added product, studies to determine fermentation properties will be needed.

The high pullulan levels elaborated in this study may be attributed to the modified Ueda medium. Since the wild type used in this study has been examined extensively [42], the ingredients would most likely affect pullulan levels since other growth conditions such as temperature, inoculation volume, media volume, and rotation speed are all consistent with other studies [12, 45, 46]. For example, sucrose has been shown in 2.5% (w/v) to allow for increased pullulan secretion and this study used 10% (w/v) sucrose [44]. Further, the nitrogen source $(NH_4)_2SO_4$, potassium source K_2HPO_4, and initial pH of 6.5 were shown to be most effective at inducing pullulan secretion, which are part of the Ueda ingredients [5, 37]. In addition, Ma et al. [33] found that the disruption of the pullulan synthetase gene results in pullulan biosynthesis ceasing. The results obtained in this study directly support this.

Comparison of all the three pullulan assays confirms pullulan production by the wild-type strain and DNA transformants, as well as a lack of production by the negative controls. Although researchers have been able to identify enzymes produced by the fungus, the pullulan biosynthesis mechanisms have still eluded researchers [31].

Proteomics can be further used to identify new enzymes produced by filamentous fungi [16]. In addition, bioinformatics is just now being used to help understand metabolic processes once hypothesized in fungi as bioinformatics was often ignored [16].

9.12 Conclusion

The goal of identifying a synthetase gene responsible for pullulan elaboration appears to have been accomplished by utilizing a 1500 bp band obtained by using an RAPD primer in PCR and transforming it into an expression vector via electroporation into *S. cerevisiae.* Transformants secreting polysaccharides and appropriate controls were then subjected to filter weights, a pullulanase assay, and the neocuproine assay to confirm pullulan elaboration. Reverse transcriptase PCR, microscopy, and genetic analysis were then also used to confirm a pullulan synthetase gene. However, the second goal of disrupting the melanin gene while the cell is concurrently producing pullulan was not accomplished in this study.

Saccharomyces expressing a pullulan synthetase gene can give researchers and the industry a new potential method for mass producing pullulan, thus making pullulan more economically feasible to utilize in many different products. Even though the melanin will still need to be bleached or removed, special media will not need to be used to grow the microbe and it could eventually even be a beneficial biproduct from other industries such as fermentation.

Acknowledgments

This project was funded by the South Dakota Agricultural Experiment Station and the South Dakota Corn Utilization Council. We would like to thank Dr. Catherine Carter for the use of laboratory equipment; Dr. Mike Hildreth and Dr. Thomas Cheesbrough for help with the bioimaging equipment; and Dr. Adrian Reber, Dr. David Henning, Dr. Bruce Bleakley, and Dr. Harold Keller for their manuscript assistance.

References

1. Amberg, D. C., Burke, D. J., and Strathern, J. N. (2005). *Methods in Yeast Genetics: A Cold Spring Harbor Laboratory Course Manual*, 2005 ed. (Cold Spring Harbor Laboratory Press, USA.)
2. An, C., Ma, S.-J., Chang, F., Xue, W.-J. (2017). Efficient production of pullulan by *Aureobasidium pullulans* grown on mixtures of potato starch hydrolysate and sucrose, *Brazilian Journal of Microbiology*, **48**(1), pp. 180–185.
3. Arora, D., Sharma, N., Sharma, V., Abrol, V., Shankar, R., and Jaglan, S. (2016). An update on polysaccharide-based nanomaterials for antimicrobial applications, *Applied Microbiology and Biotechnology*, **100**(6), pp. 2603–2615.
4. Bach, M. L., Lacroute, F., and Botstein, D. (2006). Evidence for transcriptional regulation of orotidine-5′-phosphate decarboxylase in yeast by hybridization of mRNA to the yeast structural gene cloned in *Escherichia coli*, *Proceedings of the National Academy of Sciences*, **76**(1), pp. 386–390.
5. Badr-Eldin, S. M., El-Tayeb, O. M., El-Masry, H. G., Mohamad, F. H. A., and El-Rahman, O. A. A. (1994). Polysaccharide production by *Aureobasidium pullulans*: Factors affecting polysaccharide formation. *World Journal of Microbiology & Biotechnology*, **10**, pp. 423–426.
6. Bender, H., Lehmann, J., and Wallenfels, K. (1959). Pullulan, an extracellular glucan from *Pullularia pullulans*, *Biochem. Biophys. Acta*, **36**, pp. 309–316.
7. Boeke, J. D., LaCroute, F., and Fink, G. R. (1984). A positive selection for mutants lacking orotidine-5′-phosphate decarboxylase activity in yeast: 5-fluoro-orotic acid resistance, *Molecular & General Genetics*, **197**(2), pp. 345–346.
8. Bonzi, G., Salmaso, S., Scomparin, A., Eldar-Boock, A., Satchi-Fainaro, R., and Caliceti, P. (2015). Novel pullulan bioconjugate for selective breast cancer bone metastases treatment, *Bioconjugate Chemistry*, **26**(3), pp. 489–501.

9. Bozoudi, D. and Tsaltas, D. (2018). The multiple and versatile roles of *Aureobasidium pullulans* in the vitivinicultural sector, *Fermentation*, **4**(4), 85.
10. Breierová, E., Gregor, T., Juršíková, P., Stratilová, E., and Fišera, M. (2004). The role of pullulan and pectin in the uptake of Cd^{2+} and Ni^{2+} ions by *Aureobasidium pullulans*, *Annals of Microbiology*, **54**(3), pp. 247–255.
11. Campbell, B. S., Siddique, A.-B. M., McDougall, B. M., and Seviour, R. J. (2004). Which morphological forms of the fungus *Aureobasidium pullulans* are responsible for pullulan production? *FEMS Microbiology Letters*, **232**(2), pp. 225–228.
12. Catley, B. J. (1971). Utilization of carbon sources by *Pullularia pullulans* for the elaboration of extracellular polysaccharides, *Applied Microbiology*, **22**(4), pp. 641–649.
13. Cheng, K., Demirci, A., and Catchmark, J. M. (2011). Pullulan: Biosynthesis, production, and applications, *Applied Microbiology and Biotechnology*, **92**(1), pp. 29–44.
14. Cinay, G. E., Erkoc, P., Alipour, M., Hashimoto, Y., Sasaki, Y., Akiyoshi, K., and Kizilel, S. (2017). Nanogel-integrated ph-responsive composite hydrogels for controlled drug delivery, *ACS Biomaterials Science and Engineering*, **3**(3), pp. 370–380.
15. Cullen, D., Yang, V., Jeffries, T., Bolduc, J., and Andrews, J. H. (1991). Genetic transformation of *Aureobasidium pullulans*, *Journal of Biotechnology*, **21**(3), pp. 283–288.
16. De Oliveira, J. M. P. F. and De Graaff, L. H. (2011). Proteomics of industrial fungi: Trends and insights for biotechnology, *Applied Microbiology and Biotechnology*, **89**(2), pp. 225–237.
17. Dygert, S., Li, L. H., Florida, D., and Thoma, J. A. (1965). Determination of reducing sugar with improved precision, *Analytical Biochemistry*, **13**, pp. 367–374.
18. Feng, W. Y., Austin, T. J., Chew, F., Gronert, S., and Wu, W. (2000). The mechanism of orotidine 5′-monophosphate decarboxylase: Catalysis by destabilization of the substrate, *Biochemistry*, **39**(7), pp. 1778–1783.
19. Finkelman, M. A. J., Zajic, J. E., and Vardanis, A. (1980). New method of producing protoplasts of *Aureobasidium pullulans*, *Applied and Environmental Microbiology*, **39**(4), pp. 923–925.
20. Gostinčar, C., Ohm, R. A., Kogej, T., Sonjak, S., Turk, M., Zajc, J., Zalar, P., Grube, M., Sun, H., Han, J., Sharma, A., Chiniquy, J., Ngan, C. Y., Lipzen,

A., Barry, K., Grigoriev, I. V., and Gunde-Cimerman, N. (2014). Genome sequencing of four *Aureobasidium pullulans* varieties: Biotechnological potential, stress tolerance, and description of new species, *BMC Genomics*, **15**(1), 549.

21. Henry, N., Clouet, J., Fragale, A., Griveau, L., Chédeville, C., Veziers, J., Weiss, P., Le Bideau, J., Guicheux, J., and Le Visage, C. (2017). Pullulan microbeads/Si-HPMC hydrogel injectable system for the sustained delivery of GDF-5 and TGF-B1: New insight into intervertebral disc regenerative medicine, *Drug Delivery*, **24**(1), pp. 999–1010.
22. Hong, L., Kim, W.-S., Lee, S.-M., Kang, S.-K., Choi, Y.-J., and Cho, C.-S. (2019). Pullulan nanoparticles as prebiotics enhance the antibacterial properties of *Lactobacillus plantarum* through the induction of mild stress in probiotics, *Frontier in Microbiology*, **10**, 142.
23. Israilides, C. (1994). Characterization of pullulans produced from agro-industrial wastes, *Carbohydrate Polymers*, **25**(3), pp. 203–209.
24. Israilides, C., Bocking, M., Smith, A., and Scanlon, B. (1994). A novel rapid coupled enzyme assay for the estimation of pullulan, *Biotechnology and Applied Biochemistry*, **19**(3), pp. 285–291.
25. Jund, R. and Lacroute, F. (1972). Regulation of orotidylic acid pyrophosphorylase in *Saccharomyces cerevisiae*, *Journal of Bacteriology*, **109**(1), pp. 196–202.
26. Kraśniewska, K., Ścibisz, I., Gniewosz, M., Mitek, M., Pobiega, K., and Cendrowski, A. (2017). Effect of pullulan coating on postharvest quality and shelf-life of highbush blueberry (*Vaccinium corymbosum* L.), *Materials*, **10**, 965.
27. Leathers, T. D., Nofsinger, G. W., Kurtzman, C. P., and Bothast, R. J. (1988). Pullulan production by color variant strains of *Aureobasidium pullulans*, *Journal of Industrial Microbiology*, **3**(4), pp. 231–239.
28. Leung, V., Mapletoft, J., Zhang, A., Lee, A., Vahedi, F., Chew, M., Szewczyk, A., Jahanshahi-Anbuhi, S., Ang, J., Cowbrough, B., Miller, M. S., Ashkar, A., and Filipe, C. D. M. (2019). Thermal stabilization of viral vaccines in low-cost sugar films, *Scientific Reports*, **9**, 7631.
29. Leung, V., Szewczyk, A., Chau, J., Hosseinidoust, Z., Groves, L., Hawsawi, H., Anany, H., Griffiths, M. W., Ali, M. M., and Filipe, C. D. M. (2018). Long-term preservation of bacteriophage antimicrobials using sugar glasses, *ACS Biomaterials Science and Engineering*, **4**(11), pp, 3802–3808.
30. Li, G. (1997). Genetic study of *Aureobasidium pullulans* using mutagenesis and transformation to probe for pullulan genes. Master's Degree Thesis, South Dakota State University.

31. Li, Y., Chi, Z., Wang, G.-Y., Wang, Z.-P., Liu, G.-L., Lee, C.-F., Ma, Z. C., and Chi, Z.-M. (2015). Taxonomy of *Aureobasidium spp.* and biosynthesis and regulation of their extracellular polymers, *Critical Reviews in Microbiology*, **41**(2), pp. 228–237.

32. Lieberman, I., Kornberg, A., and Simms, S. (1955). Enzymatic synthesis of pyrimidine nucleotides orotidine-5′-phosphate and uridine-5′-phosphate, *The Journal of Biological Chemistry*, **215**, pp. 403–415.

33. Ma, Z. C., Fu, W. J., Liu, G. L., Wang, Z. P., and Chi, Z. M. (2014). High-level pullulan production by *Aureobasidium pullulans* var. *melanogenium* P16 isolated from mangrove system, *Applied Microbiology and Biotechnology*, **98**(11), pp. 4865–4873.

34. Oller, A. R. (2005). *Aureobasidium pullulans* morphology: Two adapted polysaccharide stains, *Canadian Journal of Microbiology*, **51**(12), pp. 1057–1060.

35. Peterson, S. W., Manitchotpisit, P., and Leathers, T. D. (2013). *Aureobasidium thailandense* sp. nov. isolated from leaves and wooden surfaces, *International Journal of Systematic and Evolutionary Microbiology*, **63**, pp. 790–795.

36. Sheng, L., Zhu, G., and Tong, Q. (2014). Effect of uracil on pullulan production by *Aureobasidium pullulans* CGMCC1234, *Carbohydrate Polymers*, **101**(1), pp. 435–437.

37. Simon. L., Bouchet, B., Caye-Vaugien, C., and Gallant, D. J. (1995). Pullulan elaboration and differentiation of the resting forms in *Aureobasidium pullulans*, *Canadian Journal of Microbiology*, **40**, pp. 35–45.

38. Singh, R. S., Kaur, N., and Kennedy, J. F. (2015). Pullulan and pullulan derivatives as promising biomolecules for drug and gene targeting, *Carbohydrate Polymers*, **123**, pp. 190–207.

39. Thornewell, S. J., Peery, R. B., and Skatrud, P. L. (1995). Integrative and replicative genetic transformation of *Aureobasidium pullulans*, *Current Genetics*, **29**, pp. 66–72.

40. Unalan, I. U., Boyacı, D., Trabattoni, S., Tavazzi, S., and Farris, S. (2017). Transparent pullulan/mica nanocomposite coatings with outstanding oxygen barrier properties, *Nanomaterials*, **7**(281), doi: 10.3390/nano7090281.

41. Urzì, C., De Leo, F., Lo Passo, C., and Criseo, G. (1999). Intra-specific diversity of *Aureobasidium pullulans* strains isolated from rocks and other habitats assessed by physiological methods and by random amplified polymorphic DNA (RAPD), *Journal of Microbiological Methods*, **36**, pp. 95–105.

42. van Nieuwenhuijzen, E. J., Houbraken, J. A. M. P., Meijer, M., Adan, O. C. G., and Samson, R. A. (2016). *Aureobasidium melanogenum*: A native of dark biofinishes on oil treated wood. *Antonie van Leeuwenhoek, International Journal of General and Molecular Microbiology*, **109**(5), pp. 661–683.
43. Wachowska, U. and Głowacka, K. (2014). Antagonistic interactions between *Aureobasidium pullulans* and *Fusarium culmorum*, a fungal pathogen of winter wheat, *BioControl*, **59**(5), pp. 635–645.
44. West, T. P. (2011). Pullulan production by *Aureobasidium pullulans* cells immobilized in chitosan beads, *Folia Microbiologica*, **56**(4), pp. 335–338.
45. West, T. P. and Reed-Hamer, B. (1994). Elevated polysaccharide production by mutants of the fungus *Aureobasidium pullulans, FEMS Microbiology Letters*, **124**, pp. 167–172.
46. West, T. P. and Strohfus, B. (1996). Pullulan production by *Aureobasidium pullulans* grown on ethanol stillage as a nitrogen source, *Microbios*, **88**(354), pp. 7–18.
47. West, T. P. (2012). Pullulan production by *Aureobasidium pullulans* ATCC 201253 cells adsorbed onto cellulose anion and cation exchangers, *ISRN Microbiology*, **2012**, pp. 1–4.
48. www.promega.com, accessed June 1, 2019.
49. www.sigma-aldrich.com, accessed May 21, 2019.

Chapter 10

Role of Pullulans in Cosmetics

Sudipta Roy,[a] Sanjoy Kumar Das,[b] Soumalya Chakraborty,[b,c] Kamalendu Pandey,[b] Ahana Mukherjee,[d] and Rajan Rajabalaya[e]
[a]*Bengal School of Technology, Chinsurah, Hoogly 712102, West Bengal, India*
[b]*Institute of Pharmacy, Government of West Bengal, Jalpaiguri 735101, West Bengal, India*
[c]*National Institute of Pharmaceutical Education and Research (NIPER), Sector 67, S.A.S. Nagar, Mohali 160062, Punjab, India*
[d]*School of Pharmaceutical Education & Research, Jamia Hamdard (Hamdard University), Dr. Ambedkar Nagar, Delhi 110062, India*
[e]*PAPRSB Institute of Health Sciences, Universiti Brunei Darussalam, Gadong BE1410, Brunei Darussalam*
sudiptaroy1229@gmail.com

Pullulan is an anaerobically produced, water-soluble glucan gum from a yeast-like fungus *Aureobasidium pullulans*. It is a linear homopolysaccharide of glucose, which is 1,6-linked maltotriose. Pullulan has a wide range of applications commercially as well as in industries such as health care, food science, and pharmacy. Properties such as excellent transparent film-forming ability, moisture absorptivity, water solubility, and nontoxicity have made pullulan advantageous as an ingredient in the manufacturing

Pullulan: Processing, Properties, and Applications
Edited by Shakeel Ahmed and Aisverya Soundararajan
Copyright © 2021 Jenny Stanford Publishing Pte. Ltd.
ISBN 978-981-4877-17-6 (Hardcover), 978-1-003-05694-2 (eBook)
www.jennystanford.com

of cosmetics. Due to its high water solubility, it serves as an ideal ingredient for eye, bathing, and hair products. Previous works reveal that high-purity algae extract from microalgae (*Nannochloropsis oculata*) combined with pullulan has shown to tighten skin. Moreover, the presence of vitamin C and B12 as antioxidants has shown the ability to retain moisture and thus prevent water loss. Adherence properties of pullulans have instant skin-tightening effect and can thus be used to formulate antiaging creams. Argireline combined with pullulan stimulates collagen synthesis and renewal of tissue to a greater extent than Argireline alone and thus provides treatment to wrinkles. Effective binding properties of pullulans in the presence of surfactants has made it suitable for formulating creams, lotions, lipstick, lip gloss, etc. Pullulan content in a cosmetic may vary depending on its kind. Thus, pullulan may be freely used in combination with other polymers, low-molecular-weight compounds, inorganic compounds, perfumes, preservatives, etc., which are used at the time of preparation of the cosmetics, and may be used in cosmetics of any composition.

10.1 Introduction

In the Greek language, we can find a word kosmētikos, which means to "adorn" or "technique of dress and ornament." The word cosmetics actually comes from this Greek word, which is used in our daily life in a much broader sense and ways. In simpler terms, we can say that cosmetics are the substances used to decorate the human body by bringing them in contact of various parts such as hair, skin, lips, eyes, nails, etc. Starting from the ancient times, when humans used various directly obtained natural objects such as flowers, stone dust, even various parts of the animal body to decorate themselves, the journey of cosmetics to the modernization involves the alteration of the meaning and sense of cosmetics. Currently, cosmetics are not only bound to be an aid for decoration. The US Food and Drug Administration (FDA) defines the term cosmetics as a substance intended to be applied on the human body for cleansing, beautifying, and promoting attractiveness or altering the appearance of the body without hampering the body's structure and functions. It includes a cleansing agent, moisturizing agent, perfumes, sunscreen lotion, antiaging cream, powders, lotions, lipsticks, nail polishes, eye and face makeup, deodorants, baby products, hair colorants, and sprays. A wide range of acne products are developed going beyond the traditional definition of cosmetics with increased popularity among the young generation. The dark circle removing creams intended to be used by women and persons in their late forties also come under this category. The FDA sincerely makes soaps not to be considered as cosmetics [1].

Different polymers are used in cosmetics preparations; among them polymers derived from a fungal source, specifically pullulan, have received increased attention of researchers [2]. Pullulan is carbohydrate derived, anaerobically produced, water-soluble glucan gum, obtained from a fungus *Aureobasidium pullulans.* It is a linearly structured homopolysaccharide of glucopyranose, alternatively can be called a 1,6-linked maltotriose [3]. Since 1976, the Nagase Group, previously known as the Hayashibara Company (Okayama, Japan), has been the worldwide commercial producer of pullulan [4]. In 1982, the company introduced pullulan films. Currently, the Nagase Group has given permission to Pfizer for the commercialization and marketing of film-based pullulan oral care product under the name of Listerine in the United States and Canada [5]. Pullulan has a wide range of applications commercially as well as in industries such as health care, food science, pharmacy, etc. Different desired characteristics of pullulan, such as transparency, film-forming ability, moisture absorption capacity, aqueous solubility, biodegradability, and nontoxicity, have made pullulan a promising candidate in the manufacturing of cosmetics [6]. Pullulan has multiple uses in different kinds of cosmetic products such as cosmetic lotions, cosmetic powders, rouges, cosmetics for use around the eyes, facial packs, shampoos, set lotions, hair lacquers, and toothpaste. In the United States, the FDA approved pullulan as safe in 2002 [7]. According to US Patents, a cosmetic containing pullulan usually possesses a molecular weight of 5,000 to 5,000,000, and the content of total pullulan varies in the range of 0.01 to 99 parts by weight per 100 parts by weight of the total cosmetics [8]. Owing to its high water solubility, pullulan creates a luxurious smoothness for skincare and haircare products, imparting a silicone-like natural-looking after feel to skin. Pullulan also contributes in the preparation of facemasks by producing easily dissolvable films thus making it a stable and effective carrier for active ingredients, colors, and fragrances. Pullulan is also reported to be stable over a wide range of pH changes. Most importantly, pullulan does not affect the viscosity of solutions as the solution of pullulan itself is not highly viscous [9, 10]. Adherence properties of pullulans have instant skin-tightening effect and can thus be used to formulate antiaging creams. For better effect, pullulan is combined with different substances to serve cosmetic purpose like Argireline (stimulation of collagen synthesis and greater renewal of tissue), different surfactants (enhanced binding property of pullulan), high-purity algae extract from microalgae (*Nannochloropsis oculata*) (skin-tightening property), vitamin C and B12 (enhancement of moisture-retaining property), phenoxyethanol, sodium benzoate, and potassium sorbate (enhancement of tightening of skin in short and long term, protection from oxidative stress) [11], vitamin C and ferulic acid (protection against free radicals caused by environmental damage, UVA/UVB radiation and reactive oxygen species), hyaluronic acid (reduction of hyperpigmentation, erythema, and skin redness). The major disadvantages of pullulan-based cosmetics are their relatively higher cost, probably as a consequence of their expensive production. Currently, different researchers have investigated the relatively lower expensive source of carbon and nitrogen like sweet potato [12] and carob pod [13] as a medium to support the growth of *A. pullulans* [4].

10.2 Chemistry of Pullulan

The name pullulan was coined by the great scientist Bender, who described the formation of this extracellular polysaccharide by *A. pullulans* [14]. Pullulan is an anaerobically produced, water-soluble glucan gum from a yeast-like fungus *A. pullulans* [3]. Three glucose units of maltotriose are linked through α-(1→4) glycosidic bonds, whereas maltotriose units are linked through α-(1→6) bonds [15]. The chemical structure of pullulan is shown in Fig. 10.1. The molecular weight of pullulan has considerable variety, ranging from 4.5×10^4 to 6×10^5 Da [16]. The molecular weights of pullulan vary on the growing environments of the organism *A. pullulans* [17].

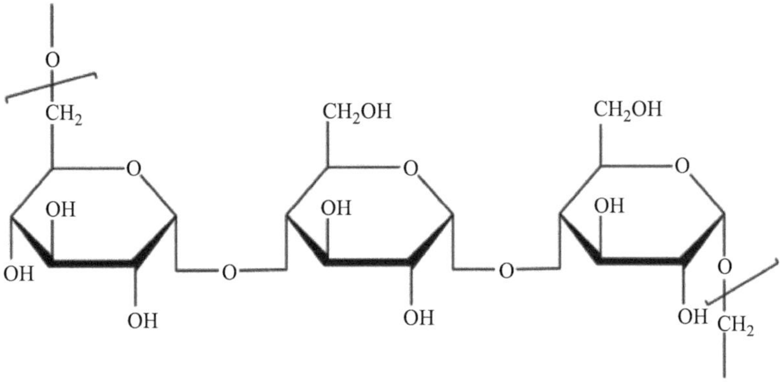

Figure 10.1 Chemical structure of pullulan. Figure taken from Ref. [18].

10.3 Solubility of Pullulan

When dissolved in hot and cold water, pullulan forms a stable, viscous solution and not a gel [16]. The study of solubility parameters led to the conclusion that pullulan is highly soluble in water and dilute alkali, whereas insoluble in alcohol and other organic solvents with an exception to dimethyl sulfoxide and formalin [2]. The viscosity of pullulan is 100–180 mm^2/s (10% w/w aqueous solution at 30°) as per the monograph [19].

10.4 Different Physicochemical Characteristics of Pullulan Useful in Imparting Cosmetic Properties

Pullulan possesses good solubility in water along with significantly low viscosity and excellent film-forming ability, which makes it a suitable candidate for the preparation of facial masks, antiwrinkle products, different haircare products such as shampoo and hair conditioner as well as shower gels. The aqueous solution of pullulan is a Newtonian fluid, is stable, and does not undergo any kind of aging or gelation. Strangely in spite of low viscosity, it possesses an excellent lubricating property. This aids to provide a soft texture to the products and a feeling of smoothness to the body after applying skincare products containing pullulan. Pullulan also possesses excellent adhesion properties, which help a cosmetic preparation containing pullulan to adhere to the skin surface, in which pullulan plays the role of the bond-forming agent. This excellent adhesion along with the relatively lower viscosity of pullulan makes pullulan-containing skincare or haircare cosmetics easy to be applied on the skin either simply by spreading by hand or mist spraying or rolling [20]. The adhesiveness of pullulan with respect to other natural polymers used in cosmetic preparation is shown in Fig. 10.2 [20]. In antiaging products containing pullulan, it provides an instant skin-tightening effect on the skin as it adheres to the skin and forms a shear film on the skin, which temporarily moderates skin texture and smoothness serving the purpose of the product [21]. After drying the aqueous solution of pullulan, a transparent film can be made. This property is exploited in the preparation of face masks and hair-styling products. Pullulan-made transparent films keep the desired style of hair for a longer period of time [22].

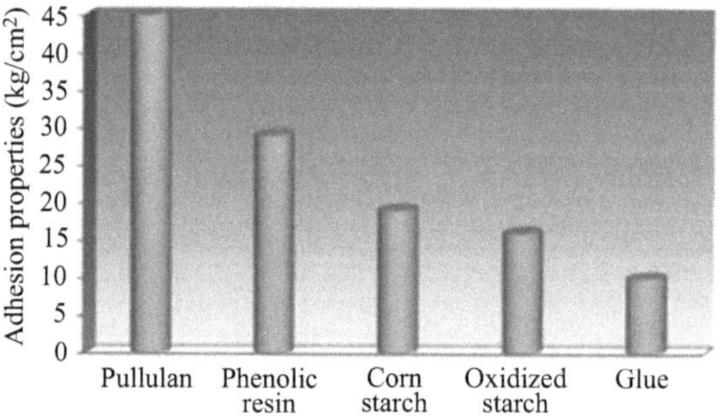

Figure 10.2 The adhesiveness of pullulan with respect to other natural polymers used in cosmetic preparation. Figure taken from Ref. [20].

10.5 Cosmetics Products Containing Pullulan

10.5.1 Antipollution Agents

Antipollution agents are one of the newest addition in the personal care and cosmetic world and have gained worldwide importance, especially among the youth. According to a survey conducted in 2016, about 79% of all the new skincare products launched throughout the world in 2016 brandished an antipollution claim. Pollution is responsible for progressive aggression of aging mainly due to its ability of degradation of the structural proteins such as collagen and elastin, resulting in a serious threat to skin firmness and elasticity [23]. It was observed by a study that an extract of microalgae *Nannochloropsis oculata*, rich in vitamin C, vitamin B12, and pullulan, provides an excellent antipollution effect and enhances skin firmness significantly. It serves the purpose of an antipollution agent by threefold mechanism [24]:

1. Supports the repair and maintenance of the extracellular membrane.
2. When applied, immediately forms a thin film on the dermal surface resulting in protection against pollution and imparting a skin-tightening effect hiding the signs of aging.
3. Stimulation of collagen formation [24].

10.5.1.1 Depolluphane as an antipollution agent

Depolluphane is an antipollution agent mainly intended to protect the skin from the effects of urban pollution. It consists of a complex containing organic cress sprout extract in a combination of a smart polysaccharide. It also provides protection to the skin cells from endogenous activator molecules, mainly due to the presence of sulforaphane, a well-known molecular activator of the cellular detoxification process of the body [25]. The main defense mechanism of Depolluphane against environmental pollutants and endogenous-activator-induced cellular damage is shown in Fig. 10.3 [25].

10.5.1.2 Depolluphane EpiPlus as an antipollution agent

Depolluphane EpiPlus is another antipollution agent with a comparatively greater spectrum of activity rather than Depolluphane. It consists of several organic extracts such as *Lepidium sativum* sprout extract and *Cynara scolymus* (Artichoke) leaf extract in combination with pullulan, sodium carboxymethyl betaglucan, *Caesalpinia spinosa* gum, maltodextrin, and water. It acts by strengthening the ability of the skin to face the stress and pollution of everyday urban lifestyle along with providing prolonged protection to the skin cells from the long-term effect of the pollution-induced epigenetic damage. The presence of the highest concentration of sulforaphane in *L. sativum* sprouts imparts a superior preventive potency against environmental pollutants as well as internal activator molecules. Depolluphane EpiPlus protects the skin from urban pollution in a threefold way:

1. It imparts a shielding effect on skin against particulate matter immediately after application.
2. It provides a short-term modulation of skin's own defense mechanism by the activation of detoxification enzymes.
3. It provides long-term protection to the dermal cells against pollutant-induced epigenetic damage [26, 27].

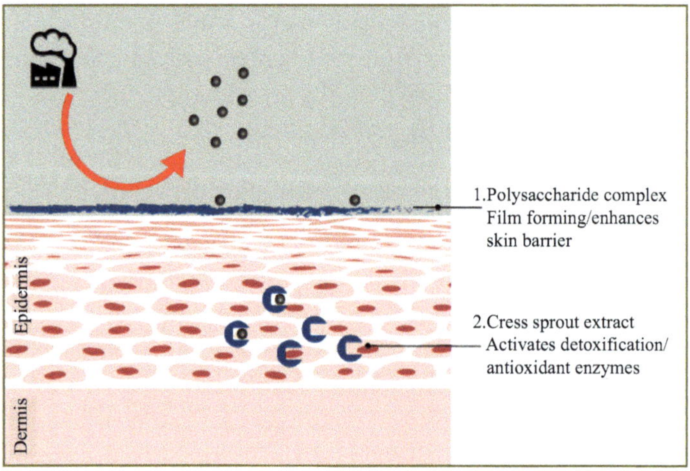

Figure 10.3 The main defense mechanism of Depolluphane against environmental pollutants and endogenous-activator-induced cellular damage [25].

10.5.1.3 PatcH2O™ as an antipollution agent

PatcH2O acts as an antipollution agent as well as a moisturizer. It consists of a micromolecular network formed by the combination of naturally occurring biopolymers along with a high concentration of moisturizing complex having glycerin, serine, trehalose, and urea as its ingredients, which prevents dermal water evaporation by acting as a protective film over the skin. The molecular mesh is also loaded with a moisturizing complex, which is gradually released into the outermost layer of skin, i.e., stratum corneum. PatcH20 is applied in antiaging and hydrating skincare formulas and also haircare products more specifically for stressed, bleached, and dry hair [28].

10.5.2 Antiwrinkle Agents

Pullulan is considered to impart a significant antiaging and antiwrinkle property to cosmetic products by the formation of a thin film, when applied on the skin surface, providing a tightening texture and appearance [21] to the skin. Cetpro pullulan, a grade of pullulan, was evaluated by manufacturers for its antiwrinkle potency, and it was observed that cetpro pullulan does not impart wrinkle reduction or whitening effects. However, when combined with vitamin C, the effects were boosted to be significantly higher than those of vitamin C alone [29].

10.5.2.1 LIFTONIN®-XPRESS as an antiwrinkle agent

LIFTONIN-XPRESS is an antiwrinkle agent formulated in combination with hydroxypropyl methylcellulose, pullulan, and *Porphyridium cruentum* extract, which acts as an instant wrinkle reducer based on the presence of skin-tightening active components (lyophilisate made from the microalga *Porphyridium cruentum* with macromolecular phycocolloids) embedded in a polymeric matrix. The antiwrinkle and skin-tightening effects are intense, long term and applicable to different kinds of dermal wrinkle. It is suitable for oil-in-water and water-in-oil emulsions, gels, and lotions [30].

10.5.3 Antioxidant

Depolluphane and Depolluphane EpiPlus (discussed in detail under antipollution agent) also possess antioxidant property mainly due to the presence of sulforaphane, which exerts a long-term protective effect against epigenetic cellular damage caused by endogenous reactive oxidizing species [25–27].

10.5.4 Binders

Pullulan, by means of its good adhesion and foam-retention properties along with its ability to form low viscous aqueous solution, can act as binder, film former, and texturing agent in several cosmetic products such as easily washable facial masks and haircare cosmetics. It is also used as a stable carrier for active ingredients, colors, and fragrances. Pullulan for this purpose is mainly produced by Nagase and is certified by ECOCERT [31].

10.5.5 Emulsifier

Kshirsagar [32] evaluated the emulsifying efficacy of pullulan in combination with gum Arabic and maltodextrin in the emulsification of turmeric oleoresin and reported the excellent emulsifying activity of pullulan supplemented with the polymeric stabilizer(s). The emulsifying activity may be due to the enhancement of viscosity of the dispersion medium by the addition of pullulan (slight enhancement alone) along with gum Arabic and maltodextrin.

10.5.5.1 JD Jojoba Aqua Cream Base

JD Jojoba Aqua Cream Base by Jojoba Desert is a light, smooth oil-in-water base moisturizing and mollifying cream. It combines the unique properties of jojoba oil with additional natural-based oils, creating a smooth ceramic texture that is slightly absorbed by the skin. It accommodates up to 20% oil, wax, silicon, water extracts, active ingredients, and more. It has a shelf life of 12 months [33].

10.5.5.2 ECOGEL™

ECOGEL is the first natural-phospholipid-based gelling-emulsifying agent, distributed in France by Lucus Meyer cosmetics and is certified by ECOCERT. It contains lysolecithin, sclerotium gum, xanthan gum along with pullulan. It increases the viscosity and the stability of formulas. It is specially designed for the formulation of gel creams. The thickening properties of ECOGEL are due to the synergistic polysaccharide combination of xanthan gum and sclerotium gum. The presence of lysolecithin provides to ECOGEL emulsifying properties with the typical phospholipid touch. The structure of lysolecithin-hydrolyzed lecithin induces a higher HLB value allowing the development of smaller oil droplets and hence more stable and thinner emulsions. Compared to classical natural gelling agents, ECOGEL provides good advantages: lower picking and resilience. Moreover, resulting textures are as sophisticated as the ones obtained with synthetic gelling agents [34].

10.5.5.3 SILIGEL™

SILIGEL is prepared by Lucas Meyer Cosmetics and finds its use as a gelling agent, co-emulsifier, stabilizer, and suspending agent. It is composed of xanthan gum, lecithin, sclerotium gum, and pullulan. With its patented synergistic combination of phospholipids and polysaccharides, this product after application provides a smooth and silicon-like feel. SILIGEL is suggested for use in various cosmetics and personal care use. It is a China-compliant ingredient [35].

10.5.6 Haircare

Pullulan is used by cosmetic manufacturers in the production of shampoo, conditioner, and different haircare products. The aqueous solution of pullulan possesses a significantly good lubricant property in spite of possessing low viscosity. This lubricating nature of the aqueous solution of pullulan is mainly responsible for its use in haircare cosmetics [36]. The recommended concentration of pullulan in different haircare cosmetics is <4% [37]. The excellent foam-promoting and builder effects are also another reason for the use of pullulan in the production of shampoo. Pullulan is also used in the manufacture of set lotions and hair lacquers due to its excellent property of tackiness and tough film-forming and hair-setting ability. Further, the property of high water solubility helps in its easy removal after use [38].

10.5.6.1 PatcH2O in haircare

PatcH2O is composed of water, glycerin, trehalose, urea, serine, pentylene glycol, glyceryl polyacrylate, algin, caprylyl glycol, sodium hyaluronate, pullulan, disodium phosphate, potassium phosphate, mainly implemented for hair cleansing, hair conditioning, hair styling, and hair treatment. This product contains a biopolymer network comprising hyaluronic acid, alginate, and pullulan, which is responsible for the formation of a protective coating on the damaged hair. PatcH2O also contains various moisturizing factors responsible for the long-lasting rapid moisture rebalancing function of the product [39]. This product is also mentioned under the antipollution agent.

10.5.7 Perfumes and Fragrances

Pullulan is used in different cosmetics in the form of easily dissolvable films to serve as a suitable and stable carrier for fragrances [40].

10.6 Toxicity Profile and Safety Considerations of Pullulan

In general, pullulan is regarded as one of the pharmaceutical ingredients with comparatively low hazardous profile. However, in 1985, a study based on *in vitro* observation of pullulan on nonmammalian cells, conducted by a team of Japanese researchers, observed that pullulan influences the occurrence of mutations, which may invite cancer [21]. The direct handling and intermittent exposure of pullulan-based cosmetics are expected to occur among hairdressers, cosmeticians, and beauticians. According to the estimation and assessment of substance exposure (EASE) (1997) model, the direct dermal exposure of pullulan-based cosmetics containing up to 10% of pullulan, in the range of 1–5 mg/cm^2/day, may lead to cancer [41], but no single component of this product at a concentration of greater than or equal to 0.1% is considered probable, possible, or confirmed human carcinogen by the International Agency for Research on Cancer (IARC) [42]. Pullulan may also produce serious systemic injury if it directly enters the bloodstream through cuts, abrasions, or lesions, so it is strictly advised not to use pullulan-based cosmetics if any external damage is present in the skin [43]. The product should be kept in a tightly closed container, should be kept away from heat, hot surfaces, sparks, open flames, and other ignition sources, and sufficient measures should be taken to prevent dust accumulation [44]. Skin irritation is observed upon exposure of a single patch containing a solution of 4% pullulan in water after 48 h [45]. The product is not considered harmful to aquatic organisms nor to cause long-term adverse effects in the environment [46]. Any respiratory allergic reaction is not reported on the use of pullulan-based cosmetics [47].

10.7 Future Challenges and Overcoming Strategies

In spite of securing a lot of captivating properties, pullulan is still getting limited usages in industrial cosmeceutical developments, mainly due to its higher production cost. So the first step for cost reduction can be satisfied by the utilization of economic resources, like by-products and wastes. Though many strategies are recently being made up of in this way, some adversities seem to take place and to make the direction very onerous, such as maintaining the supply of qualitatively and quantitatively analyzed substrates, the non-identical nutrient components of the unprocessed stuffs, and the presence of adulterated and built-up chemically inert constituents [48]. The downstream processing of fermentation broth, which mainly focuses on the removal of the pigment melanin, is stating another big problem dealing with the high rise of price. Therefore, dealing with these limitations with proper inexpensive and ecofriendly strategies and planning can lead to an increased commercially emanating usage of pullulan in the field of cosmeceuticals with its various applications [49].

Acknowledgments

The authors sincerely thank Prof. (Dr.) Jasmina Khanam, Department of Pharmaceutical Technology, Jadavpur University, for her valuable advice for this chapter.

References

1. Sharma, G. K., Gadiya, J., and Dhanawat, M. (2018). *Textbook of Cosmetic Formulations* (Pothi), ISBN: 9781365355912. https://pothi.com/pothi/book/ebook-gaurav-kumar-sharma-textbook-cosmetic-formulations (accessed on 07.08.2019).
2. Kumar, D., Saini, N., Pandit, V., and Ali, S. (2012). An insight to pullulan: A biopolymer in pharmaceutical approaches, *Int. J. Basic Appl. Sci.*, **1**(3), pp. 202–219.
3. Prajapati, V. D., Jani, G. K., and Khanda, S. M. (2013). Pullulan: An exopolysaccharide and its various applications, *Carbohydr. Polym.*, **95**, pp. 540–549.

4. Trinetta, V. and Cutter, C. N. (2016). Pullulan: A suitable biopolymer for antimicrobial food packaging applications. In: *Antimicrobial Food Packaging* (Elsevier Inc.), pp. 385–397. https://doi.org/10.1016/B978-0-12-800723-5.00030-9.
5. Trinetta, V. (2017). The application of edible and active pullulan coating on foods. In: *Reference Module in Food Science*. https://doi.org/10.1016/B978-0-08-100596-5.21129-3 (accessed on 25.08.2019).
6. LeDuy, A., Choplin, L., Zajic, J. E., and Luong, J. H. T. (2014). Pullulan: Properties, synthesis, and applications. In: *Encyclopedia of Polymer Science and Technology*, 1–14. https://doi.org/10.1002/0471440264.pst620.
7. Jindal, N. and Khattar, J. S. (2018). Microbial polysaccharides in food industry. In: *Biopolymers for Food Design*. http://dx.doi.org/10.1016/B978-0-12-811449-0.00004-9.
8. Nakashio, S., Tsuji, K., Toyota, N., and Fujita, F. (1976). Novel cosmetics containing pullulan, Patent No. 3,972,997, August 3, 1976.
9. https://www.ulprospector.com/en/eu/PersonalCare/Detail/24109/558237/Pullulan (accessed on 2.8.19).
10. Satyanarayana, T., Johri, B. N., and Prakash, A. (2012). *Microorganism in Sustainable Agriculture and Biotechnology* (Springer Netherlands). DOI:10.1007/978-94-007-2214-9.
11. https://www.makingcosmetics.com/SkinTight-AP_p_25.html (accessed on 06.08.2019).
12. Wu, S., Jin, Z., Tong, Q., and Chen, H. (2009). Sweet potato: A novel substrate for pullulan production by *Aureobasidium pullulans*. *Carbohydr. Polymer.*, **76**, pp. 645–649.
13. Roukas, T. and Biliaderis, C. G. (1995). Evaluation of carob pod as a substrate for pullulan production by *Aureobasidium pullulans. Appl. Biochem. Biotechnol.*, **55**, pp. 27–44.
14. Stankovic, I. (2019). Chemical and Technical Assessment 65th JECFA 1 PULLULAN Chemical and Technical Assessment (CTA), http://www.fao.org/fileadmin/templates/agns/pdf/jecfa/cta/65/pullulan.pdf (accessed on 2.8.19).
15. Jindal, N. and Khattar, J. S. (2018). Microbial polysaccharides in food industry. In: *Biopolymers for Food Design*. http://dx.doi.org/10.1016/B978-0-12-811449-0.00004-9.

16. Cheng, K. C., Ali, D., and Catchmark, J. M. (2011). Pullulan: Biosynthesis, production, and applications, *Appl. Microbiol. Biotechnol.*, **92**, pp. 29–44. DOI:10.1007/s00253-011-3477-y.

17. Rekha, M. R. and Sharma, C. P. (2007). Pullulan as a promising biomaterial for biomedical applications: A perspective trends biomater, *Artif. Organs*, **20**(2), pp. 111–116.

18. https://en.wikipedia.org/wiki/File:Pullulan.png.

19. Pullulan. (2005). New specifications prepared at the 65th JECFA and published in FNP 52 Add 13. An ADI 'not specified' was established was established at the 65th JECFA.

20. https://www.in-cosmetics.com/__novadocuments/216894?v=635912093969330O (accessed on 05.08.2019).

21. Pullulan, Truth in Aging. (2019) https://www.truthinaging.com/ingredients/pullulan.

22. https://korea.incosmetics.com/_novadocuments/355481?v=636289562929730000.

23. Mistry, N. (2017). Guidelines for formulating anti-pollution products, *Cosmetics*, **4**, 57. DOI:10.3390/cosmetics4040057.

24. Guillerme, J. B., Couteau, C., and Coiffard, L. (2017). Applications for marine resources in cosmetics, *Cosmetics*, **4**, 35.

25. Mibelle AG Biochemistry, Stand M20. (2019). Two lines of defense against urban pollution. https://mibellebiochemistry.com/depolluphane (accessed on 09.08.2019).

26. Depolluphane EpiPlus: Triple Action to Fight Urban Pollution (2019). Mibelle AG Biochemistry, https://www.cosmeticsandtoiletries.com/formulating/category/skincare/Depolluphane-EpiPlus-Triple-action-to-fight-urban-pollution-505082992.html.

27. https://cosmetics.specialchem.com/product/i-mibelle-biochemistry-depolluphane-epiplus.

28. https://www.ulprospector.com/en/asia/PersonalCare/Detail/1133/226311/PatcH2O.

29. www.protecingredia.com (accessed on 09.08.2019).

30. https://cosmetics.specialchem.com/product/i-rahn-liftoninxpress#frmaccesshiddeninfo (accessed on 09.08.2019).

31. https://cosmetics.specialchem.com/product/i-nagase-pullulan (accessed on 09.08.2019).

32. Kshirsagar, A. (2007). Efficacy of pullulan in emulsification of turmeric oleoresin and its subsequent microencapsulation, *XVth International Workshop on Bioencapsulation*, Vienna, Au. Sept 6-8, 2007.
33. https://cosmetics.specialchem.com/product/i-jojoba-desert-jd-jojoba-aqua-cream-base (accessed on 09.08.2019).
34. https://cosmetics.specialchem.com/product/i-lucas-meyer-cosmetics-ecogel (accessed on 09.08.2019).
35. https://cosmetics.specialchem.com/product/i-lucas-meyer-cosmetics-siligel (accessed on 09.08.2019).
36. https://korea.in-cosmetics.com/__novadocuments/355481?v=636289562929730000 (accessed on 10.08.2019).
37. https://www.ams.usda.gov/sites/default/files/media/PullulanPetition18131.pdf (accessed on 11.08.2019).
38. https://patents.google.com/patent/US3972997A/en.
39. https://www.happi.com/issues/2017-12 01/view features/new-shampoo-conditioner-ingredients (accessed on 12.08.2019).
40. https://www.ulprospector.com/en/eu/PersonalCare/Detail/24109/558237/Pullulan.
41. Pullulan, National Industrial Chemicals Notification and Assessment Scheme (NICNAS), File No. PLC/693 (2007). Hayashibara International Australia Pty Ltd, Sydney NSW 2000.
42. Pullulan, Safety Data Sheet, Carbo Synth, According to Regulation (EC) No 1907/2006, Annex II, as amended. Commission Regulation (EU) No 2015/830, https://static.cymitquimica.com/products/3D/pdf/sds-YP07957.pdf (accessed on 09.08.2019)
43. Pullulan from Aureobasidium pullulans, Santa Cruz Biotechnology, Inc., ChemWatch, sc-222228, 800.457.3801 or 831.457.3800, http://datasheets.scbt.com/sc-222228.pdf (accessed on 09.08.2019).
44. Pullulan Poly Saccharide Standard, Agilent Technologies, Inc., 800-227-9770, https://www.agilent.com/cs/library/msds/Pullulan%20Poly%20Saccharide%20Standard_NAEnglish.pdf (accessed on 09.08.2019).
45. EmulsiGel Eco. (2019) Aromantic Natural Skin Care, Aromantic Ltd,https://www.aromantic.co.uk/technical-documents/msds/emulsigel-eco-msds-aro.aspx (accessed on 09.08.2019).
46. AZCL-Pullulan, (2019) Megazyme, according to Regulation (EU) 2015/830, accessed on https://secure.megazyme.com/files/Safety_Data_Sheet/I-AZPUL_SDS.pdf (accessed on 09.08.2019).

47. Gustav Parmentier GmbH, Material Safety Data Sheet acc. REACH, DirectiveEU 453/2010 (20.05.2010), http://www.parmentier.de/gpfneu/MSDS_Pullulan_Capsules.pdf (accessed on 09.08.2019).

48. Farris, S., Introzzi, L., Fuentes-Alventosa, J. M., Santo, N., Rocca, R., and Piergiovanni, L. (2012). Self-assembled pullulan–silica oxygen barrier hybrid coatings for food packaging applications, *J. Agric. Food Chem.* **60**, pp. 782–790.

49. Singh, R. S., Saini, G. K., and Kennedy, J. F. (2008). Pullulan: Microbial sources, production and applications, *Carbohydr. Polym.*, **73**(4), pp. 515–531. DOI:10.1016/j.carbpol.2008.01.003. Epub 2008 Jan 12.

Chapter 11

Biomedical Applications of Pullulan

J. Hemapriya,[a] Ashwini Ravi,[b] Aisverya Soundararajan,[c] P. N. Sudha,[c] and S. Vijayanand[b]

[a]*PG & Research Department of Microbiology, DKM College for Women (Autonomous), Vellore 632001, Tamil Nadu, India*
[b]*Bioresource Technology Lab, Department of Biotechnology, Thiruvalluvar University, Serkkadu, Vellore 632115, Tamil Nadu, India*
[c]*Biomaterial Research Lab, Department of Chemistry, DKM College for Women (Autonomous), Vellore 632001, Tamil Nadu, India*
vipni76@gmail.com

Pullulan is an imperative exopolysaccharide produced by the yeast *Aureobasidium pullulans*. It is a water-soluble glucan gum, which was first discovered by Bauer in 1938 and was named "pullulans" by Bender in 1959. It consists of copolymer with unique linkage pattern of two α-(1→4) and one α-(1→6) glycosidic bonds in maltotriose or maltotetrose repeating units. Apart from these two linkages, it is also found to have α-(1→3) linkages. The number of maltotriose or maltotetrose units varies with various species that produce pullulan. Due to its variable structural characteristics, pullulan exhibits

Pullulan: Processing, Properties, and Applications
Edited by Shakeel Ahmed and Aisverya Soundararajan
Copyright © 2021 Jenny Stanford Publishing Pte. Ltd.
ISBN 978-981-4877-17-6 (Hardcover), 978-1-003-05694-2 (eBook)
www.jennystanford.com

numerous functional advantages, which makes it worthwhile in basic research as well as a well-defined model substance. In addition to its structural advantages, it also has several biological advantages since it is nontoxic, noncarcinogenic, nonmutagenic, biodegradable, and compatible. These biological applications made pullulan to be used in various industries, for example in the food industry as dietary fiber, prebiotic, fillers in beverages and sauces, and as an excellent replacement of starch; in the pharmaceutical industry for drug targeting, gene delivery, etc.; and also in the paper and printing industries for glossiness and lithographic printing, respectively. Despite other applications of pullulan, the pharmaceutical applications have been highly concentrated due to its solubility and inevitable biological characteristics. Therefore, this chapter focuses on the biomedical applications of pullulan.

11.1 Introduction

Pullulan is a random coil of water-soluble glucan consisting of maltotriose units joined together by α-(1→4) and α-(1→6) glycosidic linkages (Fig. 11.1) [34, 53, 67]. It is an exopolysaccharide produced primarily by *Aureobasidium pullulans*. Apart from *A. pullulans*, it is also produced by other microorganisms. Pullulan has been used as a food ingredient and as a bulking agent in the pharmaceutical industry of Japan for more than 20 years [83]. The commercial production of pullulan was initiated by Hayashibara Company Limited, Japan, in 1976. They also brought pullulan films to the market in 1982 [27, 92].

Pullulan was extensively studied by Bender, Lehmann, and Wallenfels (1959), and the name "pullulan" was coined by them [11]. The readily water-soluble pullulan has several advantageous properties such as oil resistance, impermeable to oxygen, adhesive, etc. Because of these properties, it has been utilized in various food products as dietary fiber, probiotics, pastas, baked goods, beverages, sauces, mayonnaise, etc. [74, 92]. Apart from these, it has also been used in several pharmaceutical and medicinal applications. This chapter deals with the biomedical applications of pullulans.

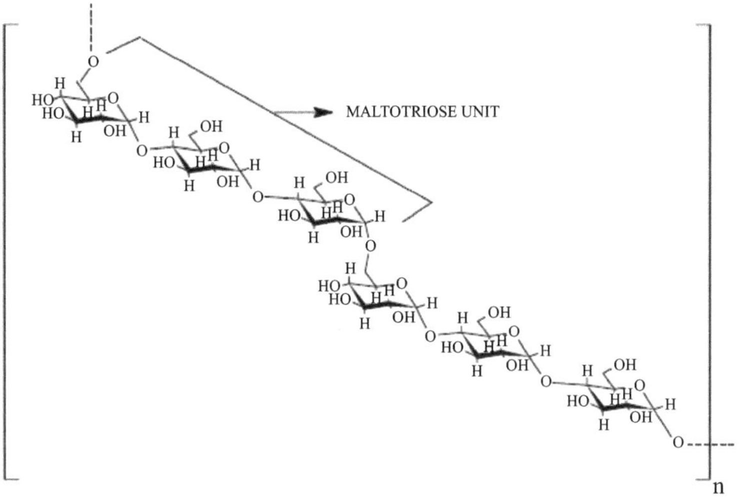

Figure 11.1 Structure of pullulan [34].

11.2 Microbial Sources of Pullulan Production

Pullulan is produced by yeast-like fungi *A. pullulans*, which is formerly called *Pulluria pullulans* [23, 49]. The microbial production of pullulan from *A. pullulans* was first discovered by Bauer in 1938 [10]. It is a polymorphic fungus, which is yeast like at the beginning of its growth and filamentous when it divides and starts producing pullulan [14, 75]. It is present in soil, water, and the saprophyte of decaying leaf, litter, wood, and other plants [13]. Apart from pullulan, *A. pullulans* also produces melanin pigment; therefore, during synthesis and production of pullulan, much care should be taken to prevent or reduce the production of melanin and also for extraction without the interference of melanin [89, 92, 95]. In addition to these, *A. pullulans* also produces various enzymes and β glucans [28, 42, 51, 52]. Apart from these, pullulan has also been produced from other species such as *Cytaria harioti* [101], *Cytaria darwinii* [73], *Cryphonectria parasitica* [25], *Teloschites flavians* [82], *Rhodototulo bacarum* [22], etc.

11.3 Properties of Pullulan

The essential advantageous properties of pullulan for which it is utilized in various food and non-food industries are listed as follows.

11.3.1 Physicochemical Properties of Pullulan

Pullulan is a water-soluble polysaccharide, which makes transparent, adhesive, colorless, viscous solution. It is insoluble in solvents such as methanol or acetone but soluble in dimethyl formamide and dimethyl sulfoxide [72, 105, 108]. It has better structural flexibility and thermal stability, which can withstand temperature of up to 250°C [35, 92]. Partial or complete esterification or etherification of pullulan make it water insoluble [11, 19, 32]. When pullulan powder is mixed with water, it forms compression moldings, membranes, films, and fibers [108].

11.3.2 Biological Properties of Pullulan

The biological properties of pullulan include biodegradability and biocompatibility. It is also edible, noncarcinogenic, nonmutagenic, nonimmunogenic, and nontoxic [30, 35, 62, 64]. Because of its structural stability, solubility, and biochemical characteristics, it has been employed widely in food and pharmaceutical applications.

11.4 Biomedical Applications of Pullulan

The physical, chemical, and biological properties of pullulan are widely applied in the pharmaceutical industry for various purposes. It has antibacterial, antifungal, antitumor, anti-inflammatory, immunomodulatory, antioxidant, antiglycemic, and antilipidemic activities [36]. Apart from these, it is utilized as drug- and gene-delivery agent, in vaccines, tissue engineering, plasma expanders, etc. These biomedical applications of pullulan are discussed as follows.

11.4.1 Drug Delivery

Pullulan has been used as a drug-delivery medium for several medical conditions. It is used either as a direct drug-delivery system or in sustainable release of drugs. The derivatives of pullulan, such as cholesterol-bearing pullulan, pullulan acetate, carboxymethyl pullulan, pullulan succinylate, and pullulan amine, have been utilized for drug release. The discontinuous epithelial cells in the lining of blood vessel in tumor tissues enable the passage of nanoparticles. This phenomenon has been utilized in the delivery of drugs. Pullulan, having excellent blood compatibility, has been utilized in drug delivery along with nanoparticles to treat tumor. In a study by Lu et al. [59], the drug Doxorubicin (DOX) was delivered using pullulan and a nanoparticle conjugate. At pH 5, 62% DOX was released in 24 h, but at pH 7.4, only 29% of DOX was released in 2 h [59]. Another study on DOX was performed by Li et al. [56], which showed that pullulan can be added with DOX of 30 wt% and nanoparticle of 50 to 110 nm. The amount of DOX conjugated with pullulan nanoparticle was found to be used as better drug-delivery system [56]. Another study on cancer treatment of pullulan along with DOX and folic acid has also been found to be effective against cancer infection [87].

The derivatives of pullulans were studied for conjugating with protein for drug delivery. The cholesteric-modified pullulan (CHP), CH-aminated pullulan (CHAP), and CH-modified carboxylated pullulan (CHSP) have been conjugated with human serum albumin (HAS) and nanoparticle. On experimenting drug release with these pullulan derivatives, CHSP–HSA had better drug-releasing activity followed by CHP–HSA [106]. Similarly, the drug Mitoxantrone (MTO), on conjugation with CHP nanoparticles and CHP–HSA nanoparticles, has been studied. The study showed that the release of the drug MTO by CHP nanoparticles was accelerated than by CHP–HSA nanoparticles at pH 5, proving that on conjugating pullulan with HSA nanoparticles, the release of drug was sustainable and persistent [107].

Pullulan has been utilized in the delivery of lipophilic drugs. Chassot et al. [18] showed that pullulan can be utilized as a polymeric stabilizer and emulsifier for lipophilic drugs by conjugating it with poly(ε-caprolactone) and beclomethasone dipropionate. The conjugate also exhibited delayed photodegradation of beclomethasone dipropionate by UVC. Pullulan acetate, a derivative of pullulan on conjugating with the drug Naproxen used for bone-related disease, showed better efficiency in drug release at variable pH [12]. In addition to these, pullulan has also been experimented to be used as oral drugs and as dry powder inhalers for lung-related disorders [33, 104].

11.4.2 Gene Delivery

Genetic diseases such as cystic fibrosis, cancer, etc. are caused due to defects in gene. The defective genes can be cured only by means of gene therapy [4, 102]. Several research on curing the defective gene without causing harmful effects are being conducted [24, 54, 84]. Several polymers have been studied to be utilized as gene-delivery systems since they are biodegradable, biocompatible, nontoxic, nonimmunogenic, etc. Among other polysaccharides, pullulan has more advantageous factors since it is an inert substance, does not cause any immunogenic activities, ready to be modified, readily soluble, and most importantly it has blood compatibility and can easily pass through the blood–brain barrier. Since it is inert, biocompatible, and does not pose any harmful effects to the host or alter the gene, it is used as an ideal carrier in gene-delivery systems for genetic therapy.

Pullulan and dextran were widely analysed for their ability to be used as gene-delivery systems. Pullulan on grafting with vinyl imidazole proved to be a better gene-delivery tool [26]. The pullulan spermine delivery system has been utilized for delivering DNA to brain endothelial cells. The addition of lipopolysaccharide to pullulan spermine delivery system still increases the efficiency of pullulan as a gene-delivery tool. Pullulan is also conjugated with protamine for gene delivery [80, 97, 98]. Dendronized pullulan with guanidine and pullulan 3D matrices with DEAE were also used as gene-delivery systems [41, 109].

11.4.3 Tissue Engineering

Tissue engineering is the process of healing damaged or lost tissues by creating a suitable cell environment for their survival, differentiation, and functional achievement. The growth factors and other nutrients have to be supplied without any interruption so that they enhance the proliferation and differentiation of cells, which in turn enhances the functionality of tissues [79, 96]. This proper supply of nutrients and other factors required for tissue engineering is provided by using an artificial structure of 3D scaffold [90]. Pullulan being a hydrophilic molecule has excellent swelling properties and forms hydrogels immediately with water [49, 94]. These hydrogels have better elasticity, porosity, and the ability to adjust with environmental and physiological fluctuations such as temperature, pH, ionic strength, polarity, surface, and mechanical properties. For these properties, pullulan has been studied for its uses in tissue engineering [9, 57].

Three-dimensional scaffolds made from pullulan gels caused wound contraction and re-epithelialization of wound cells effectively in mice [40]. Preparation of capsules using sodium alginate, pullulan, and glass ceramics showed better proliferation of osteoblasts and fibroblasts in host tissue enabling bone tissue regeneration [77]. Pullulan dextran composite and pullulan chondroitin sulfate hydrogel play an important role as hydrogels in bone tissue and cartilage tissue regeneration [29, 58]. In addition, they are also utilized in tissue engineering of injured femoral condyle bone, skull bone, and full thickness skin wound of murine models, transversal mandibular, and tibial osteotomy in goat [94].

11.4.4 Vaccination

Vaccination is essential to provoke immune response in order to prevent infectious diseases. Several vaccines came into the market for various infectious diseases. Still vaccines that can be administered by nasal and oral routes are very few, and hence several works have been performed to produce the above-mentioned vaccines for easy administration to children, ill individuals. Pullulan composites with TNF-α provoked better humoral and cell-mediated response for influenza virus [69]. On conjugation with chitosan and cationic pullulan nanogels, they were used as nasal vaccines for preventing respiratory ailments [15, 76]. Apart from their direct use as vaccine, they also play an important role in storing vaccines, especially live vaccines. Pullulan being thermally stable retained the efficacy of DNA and RNA vaccines along with trehalose [1, 55].

11.4.5 Medical Imaging

Medical imaging is the technique used for the visual interpretation of body's interior by labelling the body parts with fluorescent dyes. Quantum dots (QDs) are the extremely used imaging probe since they have broad excitation, bright fluorescence, high photostability, and narrow emission spectra. But they have the disadvantage of penetration into cells [3, 6, 17, 61, 63]. The cells do not take up QDs efficiently for fine imaging. To overcome this obstacle in imaging, QDs are now prepared with polysaccharides since they are biocompatible. Pullulan, among other polysaccharides having an advantage of blood compatibility, is produced in conjugation with QDs for better imaging [70, 78, 93].

A study was performed on pullulan nanogels acquired from cholesterol-modified pullulan and cholesterol-modified amino pullulan. From the study, it has been confirmed that cells uptake cholesterol pullulan efficiently and hence the efficiency of imaging by QDs has also been increased [71]. In a study by Chen et al., hepatic cells have been examined by contrast-enhanced ultrasound with the help of $CaCO_3$/pul-PCB (CPP) hybrid nanoparticles prepared from hepatoma-targeting pullulan [20]. Upon intravenous injection, CPP hybrids accumulated efficiently in tumor tissue and exhibited a contrast enhancement in hepatoma-bearing mice, which is six times higher than the normal imaging technique. The amino and cholesteryl group bearing pullulan is also used in PET scanning, UV–Vis scanning, and NIR fluorescence [16].

11.4.6 Plasma Expander

Efficient blood circulation and sufficient filling in the vascular system are essential for the functioning of any living being. If the volume of blood circulating or filling is reduced due to any polytrauma, blood transfusion is done [103]. Blood transfusion is performed only when there is 30% blood loss and the hemoglobin content of blood is reduced below 100 g/L. Though blood transfusion is the best solution to treat such conditions, it also has practical drawbacks such as incompatibility of blood groups, technical errors, etc. Also blood can be transfused only in finite amounts. Therefore, plasma expanders are used instead of whole blood transfusion [7]. Plasma expanders can be a crystalline or colloidal solution that can be easily stored, can be prepared plenty, and should be compatible with the system [81, 91, 99]. Pullulan being an excellent blood compatible compound has been used as plasma expander. In addition, it also does not cause any cross reaction or tissue rejection since it is biocompatible, nontoxic, and nonimmunogenic [21, 48].

11.4.7 Film-Forming Agent

The oral or nasal drugs have to be fast dissolving in order to be adsorbed easily by the body. Such fast-dissolving oral drugs are prepared especially for children, elderly, bedridden, and noncooperative patients [44]. Pullulan forms gels readily with water and can be easily dissolved. This aspect of pullulan has been exploited in drug formulations such as Class II drugs of the Biopharmaceutical Classification System [46]. Along with pullulan PEG 400, PEG 300, etc. have also been used as plasticizer in fast-dissolving drug formulations [8, 88]. For oral drugs, pullulan has been combined with a variety of drugs such as Levocetirizine dihydrochloride, Enalapril, Pilocarpine hydrochloride, Lidocaine hydrochloride, Dexamethasone, and Lactoferin [33, 44, 68].

11.4.8 Molecular Chaperones

Nanogels have been extensively used in drug- and gene-delivery systems since they can trap biomolecules, respond to external stimulus, and form macrogels in nanogel network [39, 60, 100]. These nanogels are usually prepared by synthetic methods of microemulsion, precipitation, polymerization, and intramolecular crosslinking of single-chain macromolecules [5, 31, 43, 47].

Nanogels, when conjugated with short-chain fatty acids, can bind with proteins as a host and are used as drug carriers. In addition, they can also be utilized as molecular chaperones. Since pullulan is a compatible biomolecule and forms hydrogels, it is often conjugated with nanogels to be used as molecular chaperones. Cholesterol-bearing pullulan has been extensively used as molecular chaperones for the refolding of citrate synthase [50, 85]. A study on pullulan in the thermal stabilization of horseradish peroxidase proved that pullulan can be utilized to thermally stabilize some unstable proteins [86]. In a study by Hirakura et al., spiropyrane-induced pullulan nanogels were found to have the ability to control protein folding on photostimulation [37]. Along with 2-methacryloyloxyethyl phosphorylcholine and crosslinked haluronan, pullulan is used as molecular chaperone insulin, carbonic anhydrase B and Peptide 1, erythropoietin, insulin, respectively [38, 65, 66]. Cholesterol-bearing pullulan nanogels were also found efficient in the refolding of heat-denatured enzyme carbonic anhydrase along with β cyclodextrin [2].

11.4.9 Insulinotropic Activity

The insulinotropic activity of pullulan was not much explored though a combination of sulfonylurea/pullulan along with microcapsules of islets presented higher insulin secretion level and better ability in responding to glucose changes [45].

11.5 Conclusion

Pullulan has excellent properties to be used in the medical and pharmaceutical industry as stabilizers, binders, and as gels for drug delivery, gene delivery, storing vaccines, medicinal imaging, and also in film formulations. These extensive uses of pullulan in the medical field show that pullulan has secured its place as an important polysaccharide and that production of pullulan, apart from *A. pullulans* or mutant *A. pullulans*, has to be carried out extensively to utilize it without any shortage. Still some applications of pullulan are in the beginning stage, which can be further improvised to have better and a wide variety of drugs and formulations.

References

1. A Sweet Solution for Preserving Vaccines, Science in the News, http://sitn.hms.harvard.edu/flash/2019/sweet-solution-preserving-vaccines/
2. Akiyoshi, K., Sasaki, Y., and Sunamoto, J. (1999). Molecular chaperone-like activity of hydrogel nanoparticles of hydrophobized pullulan: Thermal stabilization with refolding of carbonic anhydrase B. *Bioconjugate Chem.*, **10**(3), 321–324.
3. Alivisatos, A. P. (1996). Semiconductor clusters, nanocrystals, and quantum dots. *Science*, **271**, 933–937.
4. Anderson, W. (1992). Human gene therapy. *Science*, **256**, 808–813. doi:10.1126/science.256.5058.808.
5. Antonietti, M., Bremser, W., and Schmidt, M. (1995). Rheology of small spherical polystyrene microgels: A direct proof for a new transport mechanism in bulk polymers besides reptation. *Macromolecules*, **28**(12), 4227–4233.
6. Arya, H., Kaul Z., Wadhwa, R., Taira, K., Hirano, T., and Kaul, S. C. (2005). Quantum dots in bioimaging: Revolution by the small. *Biochem. Biophys. Res. Commun.*, **329**, 1173–1177.
7. Asskali, F. and Foster, H. (1994). Acetyl-Starke, Volumenersatz der Zukunft? *Anaesthesist*, **43**(1), 11.
8. Bala, R. and Sharma, S. (2018). Formulation optimization and evaluation of fast dissolving film of aprepitant by using design of experiment. *Bull. Fac. Pharm. Cairo Univ.*, **56**, 159–168.
9. Bataille, I. and Meddahi-pelle, A. (1982). Pullulan for biomedical use. In: *Polysaccharides in Medicinal and Pharmaceutical Applications* (Popa, V., Ed.), pp. 35–39. Smithers Rapra Technology.
10. Bauer, R. (1938). Physiology of *Dematium pullulans* de Bary. *Zentralbl Bacteriol Parasitenkd Infektionskr Hyg Abt2*, **98**, 133–167.
11. Bender, H., Lehmann, J., and Wallenfels, K. (1959). Pullulan, an extracellular glucan from *Pullularia pullulans*. *Biochimica et Biophysica Acta*, **36**, 309–316.
12. Bishwambhar, M. and Suneetha, V. (2012). Release study of Naproxen, a modern drug from pH sensitive pullulan acetate microsphere. *Int. J. Drug Dev. Res.*, **4**(4), 184–191.
13. Brand, B. G. and Kemp, H. T. (1973). Mildew defacement of organic coatings. A review of literature describing the relationship between

Aureobasidium pullulans and paint films. Paint Research Institute, Philadelphia, PA.

14. Catley, B. J. (1971). Utilization of carbon sources by *Pullularia pullulans* for the elaboration of extracellular polysaccharides. *Appl. Microbiol.*, **22**(4), 641–649.
15. Cevher, E., Salomon, S. K., Makrakis, A., Li, X. W., Brocchini, S., and Alpar, H. O. (2015). Development of chitosan–pullulan composite nanoparticles for nasal delivery of vaccines: Optimisation and cellular studies. *J. Microencapsul.*, **32**(8), 755–768.
16. Chan, M. and Almutairi, A. (2015). Nanogels as imaging agents for modalities spanning the electromagnetic spectrum. *Mater. Horiz.*, **3**, 21–40. DOI:10.1039/c5mh00161g.
17. Chan, W. C. W., Maxwell, D. J., Gao, X., Bailey, R. E., Han, M., and Nie, S. (2002). Luminescent quantum dots for multiplexed biological detection and imaging. *Curr. Opin. Biotechnol.*, **13**, 40–46.
18. Chassot, J. M., Ferreira, L. M., Gomes, F. P., Da Silva, C. D., Tasso, L., and Cruz, L. (2016). Pullulan as a stabilizer agent of polymeric nanocapsules for drug delivery. *Braz. J. Pharm. Sci.*, **52**(4), 735–740.
19. Chem. Eng. News, Japanese develop starch derived plastics (1973). 24, 40.
20. Chen, S., Xu, X. L., Zhou, B., Tian, J., Luo, B. M., and Zhang, L. M. (2019). Acidic pH-activated gas-generating nanoparticles with pullulan decorating for hepatoma-targeted ultrasound imaging. *ACS Appl. Mater. Interfaces*, **11**, 22194–22205.
21. Cheng, K. C., Demirci, A., and Catchmark, J. M. (2011). Pullulan: Biosynthesis, production, and applications. *Appl. Microbiol. Biotechnol.*, **92**(1), 29–44.
22. Chi, Z. and Zhao, S. (2003). Optimization of medium and new cultivation conditions for pullulan production by a new pullulan-producing yeast strain. *Enzyme Microb. Technol.*, **33**, 206–211.
23. Cooke, W. B. (1959). An ecological life history of *Aureobasidium pullulans* (de Bary) Arnaud. *Mycopathologia*, **12**, 1–45.
24. Cotrim, A. P. and Baum, B. J. (2008). Gene therapy: Some history, applications, problems, and prospects. *Toxicol. Pathol.*, **36**, 97–103.
25. Delben, F., Forabosco, A., Guerrini, M., Liut, G., and Torri, G. (2006). Pullulans produced by strains of *Cryphonectria parasitica*-II. Nuclear magnetic resonance evidence. *Carbohydr. Polym.*, **63**, 545–554.

26. Diana, S. M. C. and Rekha, M. R. (2017). Efficacy of vinyl imidazole grafted cationized pullulan and dextran as gene delivery vectors: A comparative study. *Int. J. Biol. Macromol.*, **105**(1), 947–955.

27. Farris, S., Unalan, I. U., Introzzi, L., Fuentes-Alventosa, J. M., and Cozzolino, C. A. (2014). Pullulan-based films and coatings for food packaging: Present applications, emerging opportunities, and future challenges. *J. App. Poly. Sci.*, **131**, Article No: 40539. DOI:10.1002/app.40539.

28. Federici, F. (1982). Extracellular enzymatic activities in *Aureobasidium pullulans*. *Mycologia*, **74**, 738–743.

29. Fricain, J. C., Schlaubitz, S., Visage, C. L., Arnault, I., Derkaoui, S. M., Siadous, R., Catros, S., Lalande, C., Bareille, R., Renard, M., Fabre, T., Cornet, S., Durand, M., Léonard, A., Sahraoui, N., Letourneur, D., and Amédée, J. (2013). A nano-hydroxyapatite–pullulan/dextran polysaccharide composite macroporous material for bone tissue engineering. *Biomaterials*, **34**, 2947–2959.

30. Fundueanu, G., Constantin, M., Mihai, D., Bortolotti, F., Cortesi, R., and Ascenzi, P. (2003). Pullulan-cyclodextrin microspheres. A chromatographic approach for the evaluation of the drug-cyclodextrin interactions and the determination of the drug release profiles. *J. Chromatogr. B*, **791**, 407–419.

31. Gan, D. and Lyon, L. A. (2001). Tunable swelling kinetics in core–shell hydrogel nanoparticles. *J. Am. Chem. Soc.*, **123**(31), 7511–7517.

32. Ger. Pat. 1,096,850 (Jan. 12, 1961), K. Wallenfels and H. Bender (to Farbwerke Hoechst AG).

33. Gherman, S., Zavastin, D., Ochiuz, L., Biliuta, G., and Coseri, S. (2016). Enalapril maleate loaded pullulan film for mucoadhesive buccal drug delivery applications. *Cellulose Chem. Technol.*, **50**(5–6), 593–600.

34. Gibbs, P. A. and Seviour, R. J. (1996). Pullulan. In: *Polysaccharides in Medicinal Applications* (Dimitiu, S., Ed.), pp. 59–86. Marcel Dekker, New York.

35. Gniewosz, M. and Duszkiewicz-Reinhard, W. (2008). Comparative studies on pullulan synthesis, melanin synthesis and morphology of white mutant *Aureobasidium pullulans* B-1 and parent strain A.p.-3. *Carbohydr. Polym.*, **72**, 431–438.

36. Grigoras, A. G. (2019). Drug delivery systems using pullulan, a biocompatible polysaccharide produced by fungal fermentation of starch. *Environ. Chem. Lett.*, **17**(3), 1209–1223.

37. Hirakura, T., Nomura, Y., Aoyama, Y., and Akiyoshi, K. (2004). Photoresponsive nanogels formed by the self-assembly of spiropyrane-bearing pullulan that act as artificial molecular chaperones. *Biomacromolecules*, **5**(5), 1804–1809.

38. Hirakura, T., Yasugi, K., Nemoto, T., Sato, M., Shimoboji, T., Aso, Y., Morimoto, N., and Akiyoshi, K. (2010). Hybrid hyaluronan hydrogel encapsulating nanogel as a protein nanocarrier: New system for sustained delivery of protein with a chaperone-like function. *J. Control. Release*, **142**, 483–489.

39. Hu, Z., Lu, X., Gao, J., and Wang, C. (2000). Polymer gel nanoparticle networks. *Adv. Mater.*, **12**(16), 1173–1176.

40. Iswariya, S., Bhanukeerthi, A. V., Velswamya, P., Uma, T. S., and Perumal, P. T. (2013). Design and development of a piscine collagen blended pullulan hydrogel for skin tissue engineering. *RSC Adv.*, **6**(63), 57863–57871.

41. Juan, A. S., Hlawaty, H., Chaubet, F., Letourneur, D., and Feldman, L. J. (2007). Cationized pullulan 3D matrices as new materials for gene transfer. *J. Biomed. Mater. Res. Part A*, **82**(22), 354–362.

42. Jung, M. Y., Kim, J. W., Kim, K. Y., Choi, S. H., and Ku, S. K. (2016). Polycan, a beta-glucan from *Aureobasidium pullulans* SM-2001, mitigates ovariectomy-induced osteoporosis in rats. *Exp. Ther. Med.*, **12**, 1251–1262.

43. Kadlubowski, S., Grobelny, J., Olejniczak, W., Cichomski, M., and Ulanski, P. (2003). Pulses of fast electrons as a tool to synthesize poly(acrylic acid) nanogels. Intramolecular cross-linking of linear polymer chains in additive-free aqueous solution. *Macromolecules*, **36**(7), 2484–2492.

44. Kathpalia, H. and Patil, A. (2017). Formulation and evaluation of orally disintegrating films of levocetirizine dihydrochloride. *Indian J. Pharm. Sci.*, **79**(2), 204–211.

45. Kim, S., Chae, S. Y., Nab, K., Kim, S. W., and Bae, Y. H. (2003). Insulinotropic activity of sulfonylurea/pullulan conjugate in rat islet microcapsule. *Biomaterials*, **24**(26), 4843–4851.

46. Krull, S. M., Ma, Z., Li, M., Dave, R. N., and Bilgili, E. (2015). Preparation and characterization of fast dissolving pullulan films containing BCS class II drug nanoparticles for bioavailability enhancement. *Drug Dev. Ind. Pharm.*, **42**(7), 1073–1085.

47. Kuckling, D., Vo, C. D., and Wohlrab, S. E. (2002). Preparation of nanogels with temperature-responsive core and pH-responsive arms by photo-cross-linking. *Langmuir*, **18**(11), 4263–4269.

48. Kulicke, W.-M. and Heinze, T. (2006). Improvements in polysaccharides for use as blood plasma expanders. *Macromol. Symp.*, **231**(1), 47–59.

49. Kumar, D., Saini, N., Pandit, V., and Ali, S. (2012). An insight to pullulan: A biopolymer in pharmaceutical approaches. *Int. J. Basic Appl. Sci.*, **1**, 202–219.

50. Kuroda, K., Fujimoto, K., Sunamoto, J., and Akiyoshi, K. (2002). Hierarchical self-assembly of hydrophobically modified pullulan in water: Gelation by networks of nanoparticles. *Langmuir*, **18**(10), 3780–3786.

51. Leathers, T. D. (1987). Host amylases and pullulan production. In: *First Materials Biotechnology Symposium. Technical Report NATICK/TR-88/033* (Kaplan, D. L., Ed.), pp. 175–185. US Army, Natick, MA.

52. Leathers, T. D. (1993). Substrate regulation and specificity of amylases from *Aureobasidium* strain NRRLY-12,974. *FEMS Microbiol. Lett.*, **110**, 217–222.

53. Leathers, T. D. (2003). Biotechnological production and applications of pullulan. *Appl. Microbiol. Biotechnol.*, **62**, 468–473.

54. Leiden, J. M. (1995). Gene therapy: Promise, pitfalls, and prognosis. *N. Engl. J. Med.*, **333**, 871–873.

55. Leung, V., Mapletoft, J., Zhang, A., Lee, A., Vahedi, F., Chew, M., Szewczyk, A., Jahanshahi-Anbuhi, S., Ang, J., Cowbrough, B., Miller, M. S., Ashkar, A., and Filipe, C. D. M. (2019). Thermal stabilization of viral vaccines in low-cost sugar films. *Sci. Rep.*, **9**, 7631.

56. Li, H., Bian, S., Huang, Y., Liang, J., Fan, Y., and Zhang, X. (2014). High drug loading pH-sensitive pullulan-DOX conjugate nanoparticles for hepatic targeting. *J. Biomed. Mater. Res. Part A*, **102A**, 150–159.

57. Li, H., Yang, J., Hu, X., Liang, J., Fan, Y., and Zhang, X. (2011). Superabsorbent polysaccharide hydrogels based on pullulan derivate as antibacterial release wound dressing. *J. Biomed. Mater. Res. A*, **98**(1), 31–39.

58. Li, T., Song, X., Weng, C., Wang, X., Sun, L., Gong, X., Yanga, L., and Chen, C. (2018). Self-crosslinking and injectable chondroitin sulfate/pullulan hydrogel for cartilage tissue engineering. *Appl. Mater. Today*, **10**, 173–183.

59. Lu, D., Wen, X., Liang, J., Gu, Z., Zhang, X., and Fan, Y. (2008). A pH-sensitive nano drug delivery system derived from pullulan/doxorubicin conjugate. *J. Biomed. Mater. Res. Part B: Appl. Biomater.*, **89**(1), 177–183.

60. McAllister, K., Sazani, P., Adam, M., Cho, M. J., Rubinstein, M., Samulski, R. J., and DeSimone, J. M. (2002). Polymeric nanogels produced via inverse microemulsion polymerization as potential gene and antisense delivery agents. *J. Am. Chem. Soc.*, **124**(51), 15198–15207.
61. Michalet, X., Pinaud, F. F., Bentolila, L. A., Tsay, J. M., Doose, S., Li, J. J., Sundaresan, G., Wu, A. M., Gambhir, S. S., and Weiss, S. (2005). Quantum dots for live cells, in vivo imaging, and diagnostics. *Science*, **307**, 538–544.
62. Mishra, B., Vuppu, S., and Rath, K. (2011). The role of microbial pullulan, a biopolymer in pharmaceutical approaches: A review. *J. Appl. Pharm. Sci.*, **01**(06), 45–50.
63. Mitchell, P. (2001). Turning the spotlight on cellular imaging. *Nat. Biotechnol.*, **19**, 1013–1017.
64. Mocanu, G., Mihai, D., Le Clerf, D., Picton, L., and Muller, G. (2004). Synthesis of new associative gel microspheres from carboxymethyl pullulan and their interactions with lysozyme. *Eur. Polym. J.*, **40**, 283–289.
65. Morimoto, N., Endo, T., Iwasaki, Y., and Akiyoshi, K. (2005a). Design of hybrid hydrogels with self-assembled nanogels as cross-linkers: Interaction with proteins and chaperone-like activity. *Biomacromolecules*, **6**, 1829–1834.
66. Morimoto, N., Endo, T., Ohtomi, M., Iwasaki, Y., and Akiyoshi, K. (2005b). Hybrid nanogels with physical and chemical cross-linking structures as nanocarriers. *Macromol. Biosci.*, **5**, 710–716.
67. Morris, V. J. (1995). Bacterial polysaccharides. In: *Food Polysaccharides and Their Applications* (Stephen, A. M., Ed.), pp. 341–375. Marcel Dekker, New York.
68. Murata, Y., Isobe, T., Kofuji, K., Nishida, N., and Kamaguchi, R. (2010). Preparation of fast dissolving films for oral dosage from natural polysaccharides. *Materials*, **3**, 4291–4299.
69. Nagatomo, D., Taniai, M., Ariyasu, H., Taniguchi, M., Aga, M., Ariyasu, T., Ohta, T., and Fukuda, S. (2015). Cholesteryl pullulan encapsulated TNF-α nanoparticles are an effective mucosal vaccine adjuvant against influenza virus. *BioMed Res Int.*, **2015**, Article ID 471468. http://dx.doi.org/10.1155/2015/471468.
70. Nobuyuki, M., Takao, E., Michiko, O., Yasuhiko, I., and Kazunari, A. (2005). Hybrid nanogels with physical and chemical cross-linking structures as nanocarriers. *Macromol. Biosci.*, **5**, 710–716.

71. Numata, M., Asai, M., Kaneko, K., Bae, A. H., Hasegawa, T., Sakurai, K., and Shinkai, S. (2005). Inclusion of cut and as-grown single-walled carbon nanotubes in the helical superstructure of schizophyllan and curdlan (N-13-glucans). *J. Am. Chem. Soc.*, **127**, 5875–5884.

72. Oguzhan, P. and Yangilar, F. (2013). Pullulan: Production and usage in food industry. *Afr. J. Food Sci. Technol.*, **4**, 57–63.

73. Oliva, E. M., Cirelli, A. F., and De Lederkremer, R. M. (1986). Characterization of a pullulan in *Cyttaria darwinii*. *Carbohydr. Res.*, **158**, 262–267.

74. Oliveira, J. D., Lopes, L. M. A., Pimenta, F. D., and Sérvulo, E. F. C. (2015). Low cost production of pullulan obtained from granulated sugar and different nitrogen source. *IJET-IJENS*, **15**(1), 11–17.

75. Ono, K., Yasuda, N., and Ueda, S. (1977). Effect of pH on pullulan elaboration by *Aureobasidium pullulans* S-lt. *Agric. Biol. Chem.*, **41**(11), 2113–2118.

76. Ouchidaa, R. K., Yukia, Y., and Kiyono, H. (2018). Cationic pullulan nanogel as a safe and effective nasal vaccine delivery system for respiratory infectious diseases. *Hum. Vacc. Immunother.*, **14**(9), 2189–2193.

77. Popescu, R. A., Magyari, K., Taulescu, M., Vulpoi, A., Berce, C., Bogdan, S., Lelescu, C., Dreancă, A., Tudoran, O., Papuc, I., and Baia, L. (2018). New alginate-pullulan-bioactive glasses composites with copper oxide for bone tissue regeneration trials, **12**(10), 2112–2121.

78. Prajapati, V. D., Jani, G. K., and Khanda, S. M. (2013). Pullulan: An exopolysaccharide and its various applications. *Carbohydr. Polym.*, **95**, 540–549.

79. Priya, S. G., Jungvid, H., and Kumar, A. (2008). Skin tissue engineering for tissue repair and regeneration. *Tissue Eng. Part B Rev.*, **14**, 105–118.

80. Priya, S. S., Rekha, M. R., and Sharma, C. P. (2013). Pullulan–protamine as efficient haemocompatible gene delivery vector: Synthesis and in vitro characterization. *Carbohydr. Polym.*, **102**(2014), 207–215.

81. Ragaller, M. J., Theilen, H., and Koch, T. (2001). Volume replacement in critically ill patients with acute renal failure. *J. Am. Soc. Nephrol.*, **12**(17), 33–39.

82. Reis, R. A., Tischer, C. A., Gorrin, P. A. J., and Iacomini, M. (2002). A new pullulan and a branched (1→3)-, (1→6)-linked b-glucan from the lichenised ascomycete *Teloschistes flavicans*. *FEMS Microbiol. Lett.*, **210**, 1–5.

83. Rekha, M. R. and Sharma, C. P. (2009). Blood compatibility and in vitro transfection studies on cationically modified pullulan for liver cell targeted gene delivery. *Biomaterials*, **30**, 6655–6664.

84. Rosenberg, S. A., Aebersold, P., Cornetta, K., Kasid, A., Morgan, R. A., Moen, R., Karson, E. M., Lotze, M. T., Yang, J. C., Topalian, S. L., Merino, M. J., Culver, K., Miller, A. D., Blaese, R. M., and Anderson, W. F. (1990). Gene transfer into humans: Immunotherapy of patients with advanced melanoma, using tumor-infiltrating lymphocytes modified by retroviral gene transduction. *N. Engl. J. Med.*, **323**, 570–578.

85. Sawada, S., Nomura, Y., Aoyama, Y., and Akiyoshi, K. (2006). Heat shock protein-like activity of a nanogel artificial chaperone for citrate synthase. *J. Bioact. Compat. Polym.*, **21**(6), 487–501.

86. Sawada, S., Sasaki, Y., Nomura, Y., and Akiyoshi, K. (2011). Cyclodextrin-responsive nanogel as an artificial chaperone for horseradish peroxidase. *Colloid Polym. Sci.*, **289**(5–6), 685–691.

87. Scomparina, A., Salmasoa, S., Bersania, S., Satchi-Fainarob, R., and Caliceti, P. (2011). Novel folated and non-folated pullulan bioconjugates for anticancer drug delivery. *Eur. J. Pharm. Sci.*, **42**(5), 547–558.

88. Seema Saini, S., Rana, A. C., and Gupta, S. (2011). Optimization of formulation of fast dissolving films made of pullulan polymer, **9**(1), 127–131.

89. Seihr, D. J. (1981). Melanin biosynthesis in *Aureobasidium pullulans*. *J. Coat. Technol.*, **53**, 23–25.

90. Shingel, K. I. (2004). Current knowledge on biosynthesis, biological activity, and chemical modification of the exopolysaccharide, pullulan. *Carbohydr. Res.*, **339**(3), 447–460.

91. Singer, P., Kogan, A., Zolotarski, V., Grozovski, E., and Cohen, J. (1999). Extracellular fluid variations during a fluid challenge: A comparison of normal saline (NS) and starch (HES) in stressed patients. *Crit. Care*, **3**(1), 152.

92. Singh, R. S., Saini, G. K., and Kennedy, J. F. (2008). Pullulan: Microbial sources, production and applications. *Carbohydr. Polym.*, **73**(4), 515–531.

93. Singh, R. S., Kaur, N., Rana, V., and Kennedy, J. F. (2017). Pullulan: A novel molecule for biomedical applications. *Carbohydr. Polym.*, **171**, 102–121.

94. Singh, R. S., Kaur, N., Rana, V., and Kennedy, J. F. (2016). Recent insights on applications of pullulan in tissue engineering. *Carbohydr. Polym.*, **153**, 455–462.

95. Singh, R. S., Saini, G. K., and Kennedy, J. F. (2009). Downstream processing and characterization of pullulan from a novel colour variant strain of *Aureobasidium pullulans* FB-1. *Carbohydr. Polym.*, **78**, 89–94.

96. Tabata, Y. (2003). Tissue regeneration based on drug delivery technology. In: *Topics in Tissue Engineering* (Ashammakhi, N. and Ferretti, P., Eds.), pp. 1–32. University of Oulu, Oulu.

97. Thomsen, L. B., Lichota, J., Kim, K. S., and Moos, T. (2011). Gene delivery by pullulan derivatives in brain capillary endothelial cells for protein secretion. *J. Control. Release*, **151**, 45–50.

98. Ueda, M., Jo, J., Gao, J. Q., and Tabata, Y. (2019). Effect of lipopolysaccharide addition on the gene transfection of spermine-introduced pullulan-plasmid DNA complexes for human mesenchymal stem cells. *J. Biomater. Sci.*, **30**(16), 1542–1558.

99. Van der Sande, F. M., Luik, A. J., Kooman, J. P., Verstappen, V., and Leunissen, M. L. (2000). Effect on intravenous fluids on blood pressure course during hemodialysis in hypotensive-prone patients. *J. Am. Soc. Nephrol.*, **11**(3), 550–555.

100. Vinogradov, S. V., Bronich, T. K., and Kabanov, A. V. (2002). Nanosized cationic hydrogels for drug delivery: Preparation, properties and interaction with cells. *Adv. Drug Delivery Rev.*, **54**(1), 135–147.

101. Waksman, N., De Lederkremer, R. M., and Cerezo, A. S. (1977). The structure of an a-D-glucan from *Cyttaria harioti* Fischer. *Carbohydr. Res.*, **59**, 505–515.

102. Why: Gene Therapy? (1997). Publ. Online, DOI:101038sjgt3300488.4.

103. Winslow, R. M. (1995). Blood substitutes, a moving target. *Nat. Med.*, **1**, 1212–1215.

104. Xu, E., Jiang, W., Xu, Y., Cai, Y., and Li, H. (2015). Pullulan-modified spray-dried powders for pulmonary drug delivery. *J. Control. Release*, **213**, e8–e152.

105. Youssef, F., Roukas, T., and Biliaderis, C. G. (1999). Pullulan production by a nonpigmented strain of *Aureobasidium pullulans* using batch and fed-batch culture. *Process Biochem.*, **34**, 355–366.

106. Yuan, L., Cao, Y., Luo, Q., Yang, W., Wu, X., Yang, X., Wu, D., Tan, S., Qin, G., Zhou, J., Zeng, Y., Chen, X., Tao, X., and Zhang, Q. (2018). Pullulan-based nanoparticle-HSA complex formation and drug release influenced by surface charge. *Nanoscale Res. Lett.*, **13**, 317.

107. Yuan, L., Guo, B., Zhong, W., Nie, Y., Yao, X., Peng, X., Wang, R., Yu, H., Yang, S., He, C., Tao, X., and Zhang, Q. (2019). Interaction of mitoxantrone-loaded cholesterol modified pullulan nanoparticles with human

serum albumin and effect on drug release. *J. Nanomater.*, **2019**, Article ID 8036863. https://doi.org/10.1155/2019/8036863.

108. Yuen, S. (1974). Pullulan and its applications. *Process Biochem.*, **22**, 17–22.

109. Zhou, J., Wali, A. R. M., Ma, S., He, Y., Yue, D., Tang, J. Z., and Gu, Z. (2018). Tailoring the supramolecular structure of guanidinylated pullulan toward enhanced genetic photodynamic therapy. *Biomacromolecules*, **19**(6), 2214–2226.

Index